Strengthening Structural Concrete with Fiber-Reinforced Polymer (FRP) Systems—Code Requirements and Commentary

An ACI Standard

Reported by ACI Committee 440S

| Kent A. Harries, Chair | Maria Lopez de Murphy, Vice Chair | William J. Gold, Secretary |

Tarek Alkhrdaji	Mahmut Ekenel	Abheetha Peiris	Consulting Member
Scott Arnold	Ravi Kanitkar	J. Gustavo Tumialan	Carl J. Larosche
Aniket Borwankar	Michael W. Lee	Erblina Vokshi	

ACI CODE-440.13 was developed to provide design professionals a code for the design of strengthening strategies for concrete structures using fiber-reinforced polymer (FRP) systems.

Keywords: buildings; carbon fiber; fiber-reinforced polymer; glass fiber; rehabilitation; repair; strengthening; structural design.

ACI CODE-440.13-24 was approved by the ACI Standards Board for publication June 5, 2024, and published September 2024.

Copyright © 2024, American Concrete Institute.

All rights reserved including rights of reproduction and use in any form or by any means, including the making of copies by any photo process, or by electronic or mechanical device, printed, written, or oral, or recording for sound or visual reproduction or for use in any knowledge or retrieval system or device, unless permission in writing is obtained from the copyright proprietors.

CONTENTS

PREFACE, p. 4

CHAPTER 1—GENERAL, p. 5
1.1—Scope 5
1.2—General 5
1.3—Purpose 5
1.4—Interpretation 6
1.5—Authority having jurisdiction 7
1.6—Licensed design professional 7
1.7—Inspector 7
1.8—Design and construction documents 7
1.9—Testing and inspection 8
1.10—Approval of special systems of design, construction, or alternative construction materials 8

CHAPTER 2—NOTATION AND DEFINITIONS, p. 9
2.1—Code notation 9
2.2—Definitions 12

CHAPTER 3—REFERENCED STANDARDS, p. 15
3.1—American Concrete Institute 15
3.2—American Society of Civil Engineers 15
3.3—ASTM International 15
3.4—Underwriters Laboratories 16

CHAPTER 4—FRP SYSTEM REQUIREMENTS, p. 17
4.1—General 17
4.2—Wet layup FRP systems 17
4.3—Precured carbon FRP systems 17
4.4—FRP near-surface-mounted bars 18

CHAPTER 5—CONCRETE SUBSTRATE REQUIREMENTS, p. 20
5.1—General 20
5.2—Bond critical externally bonded FRP systems 20
5.3—Contact critical externally bonded FRP systems ...21
5.4—NSM FRP systems 21

CHAPTER 6—GENERAL DESIGN REQUIREMENTS, p. 22
6.1—General 22
6.2—Basis of design 22
6.3—Load factors and combinations 22
6.4—Design material properties 23
6.5—Maximum sustained loads 23
6.6—Maximum service temperature 24
6.7—Durability requirements 24

CHAPTER 7—DESIGN AND DETAILING FOR FLEXURAL STRENGTHENING, p. 25
7.1—General 25
7.2—Design strength 25
7.3—Design requirements 26
7.4—Nominal flexural strength 28
7.5—Moment redistribution for continuous reinforced concrete beams 31
7.6—Anchorage and development of externally bonded FRP 31
7.7—Development of near-surface-mounted FRP flexural strengthening 33

CHAPTER 8—DESIGN AND DETAILING FOR SHEAR STRENGTHENING, p. 35
8.1—General 35
8.2—Sectional requirements 35
8.3—FRP System wrapping schemes 35
8.4—Design strength 36
8.5—Nominal shear strength 36
8.6—FRP Contribution to shear strength 36
8.7—Anchorage for U-wraps 38
8.8—Details for FRP shear strengthening 41

CHAPTER 9—DESIGN AND DETAILING FOR AXIAL FORCE AND COMBINED AXIAL FORCE AND MOMENT STRENGTHENING, p. 42
9.1—General 42
9.2—Axial compression 42
9.3—Combined axial compression and bending 46

CHAPTER 10—FIRE RESISTANCE, p. 48
10.1—General 48
10.2—Fire resistance of FRP-strengthened members 48

CHAPTER 11—FIELD INSPECTION, TESTING, AND EVALUATION, p. 50
11.1—General 50
11.2—Field inspection 50
11.3—Material testing 51
11.4—Evaluation and acceptance criteria 52
11.5—Inspection of coatings 52

COMMENTARY REFERENCES, p. 53
Authored documents 54

APPENDIX A—ADDITIONAL LOAD COMBINATIONS FOR FRP STRENGTHENING, p. 56
A.1—Appendix notation 56
A.2—Scope 56
A.3—Additional load combinations for FRP strengthening 56

PREFACE

This Code provides minimum design requirements for strengthening of existing concrete structural systems and members using externally bonded and near-surface-mounted (NSM) fiber-reinforced polymer (FRP) systems. Among the subjects covered are design and detailing for strength, serviceability, and durability; load combinations, load factors, and strength reduction factors; FRP anchorage to concrete; development and splicing of FRP reinforcement; field inspection; and testing. This Code was developed by a consensus process. This Code is written for use by licensed design professionals and authorities having jurisdiction. This Code provides minimum requirements for materials, design and construction, and quality control and assurance requirements for FRP strengthening systems. This Code is written in a format that allows adoption by reference in a repair code or a general building code. Background details or suggestions for carrying out the requirements or intent of this Code provisions are in the Commentary.

PAGE LEFT INTENTIONALLY BLANK

CODE

CHAPTER 1—GENERAL

1.1—Scope

1.1.1 ACI CODE-440.13, "Code Requirements for Strengthening Structural Concrete with Fiber-Reinforced Polymer (FRP) Systems—Code and Commentary," is hereafter referred to as "this Code." is presented in a side-by-side column format. These are two separate but coordinated documents, with Code text placed in the left column and the corresponding Commentary text aligned in the right column. Commentary section numbers are preceded by an "R" to further distinguish them from Code section numbers. The two documents are bound together solely for the user's convenience. Each document carries a separate enforceable and distinct copyright.

1.1.2 This Code shall apply to the strengthening of existing concrete structures using only those unidirectional externally bonded and near-surface-mounted (NSM) fiber-reinforced polymer (FRP) systems permitted in Chapter 4.

1.1.3 This Code provides minimum requirements for the materials, design, and construction of FRP strengthening systems for concrete structures consistent with the requirements of ACI CODE-562.

1.1.4 This Code shall not be applied to the strengthening of masonry structures.

1.1.5 This Code provides minimum requirements for the strength evaluation, testing, and inspection of FRP strengthening systems for concrete structures consistent with the requirements of ACI CODE-562.

1.2—General

1.2.1 The requirements of this Code use strength design provisions for demands and capacities.

1.2.2 FRP strengthening is permitted for the following:
(a) All members in structures assigned to Seismic Design Category (SDC) A in accordance with ASCE/SEI 7
(b) Structural members not designated as part of the seismic-force-resisting system in all SDCs.

1.3—Purpose

1.3.1 The purpose of this Code is to provide for public health and safety by establishing minimum requirements for

COMMENTARY

CHAPTER R1—GENERAL

R1.1—Scope

R1.1.2 Throughout this Code, the term "structure" means an existing building, nonbuilding structure, member, or system.

R1.1.3 This Code focuses on concrete buildings and nonbuilding structures. For buildings or structures similar to buildings, members that are addressed by this Code include concrete portions of composite members, and precast and prestressed concrete members.

The licensed design professional can perform assessment, design, and quality assurance activities that exceed the minimum requirements of this Code. Requirements beyond the minimum stated in this Code, such as those for long-term durability, redundancy, or integrity, can be considered by the licensed design professional.

R1.1.4 Guidance for strengthening of masonry structures is provided in ACI PRC-440.7.

R1.2—General

R1.2.2 Seismic strengthening of members of the seismic-force-resisting systems in structures assigned to SDC B through F is outside the scope of this Code. Other standards, such as ACI CODE-369.1 and ASCE/SEI 41, address repair and strengthening of seismic-force-resisting systems. ACI PRC-440.2 provides guidance for the use of FRP systems for strengthening seismic-force-resisting systems.

R1.3—Purpose

CODE

strength, stability, serviceability, durability, and fire resistance of concrete structures strengthened using FRP systems.

1.3.2 This Code does not address all design considerations.

1.3.3 Construction means and methods are not addressed in this Code.

1.4—Interpretation

1.4.1 The official version of this Code is the English language version using inch-pound units published by the American Concrete Institute.

1.4.2 In case of conflict between the official version of this Code and other versions of this Code, the official version governs.

1.4.3 Modifications to this Code that are adopted by a particular jurisdiction are part of the laws of that jurisdiction but are not a part of this Code.

1.4.4 If provisions in this Code conflict with the regulations governing existing structures of the authority having jurisdiction, or those of a code adopting this Code, the regulations of the authority having jurisdiction or the adopting code shall govern.

1.4.5 If provisions in this Code conflict with requirements of standards referenced within this Code, this Code shall govern.

1.4.6 This Code shall be interpreted in a manner that avoids conflict between or among its provisions. Specific provisions shall govern over general provisions.

1.4.7 The Commentary consists of a preface, commentary text, tables, figures, and cited publications. The Commentary is intended to provide contextual information, but is not part of this Code, does not provide binding requirements, and shall not be used to create a conflict with or ambiguity in this Code.

1.4.8 This Code shall be interpreted and applied in accordance with the plain meaning of the words and terms used. Specific definitions of words and terms in this Code shall be used where provided and applicable, regardless of whether other materials, standards, or resources outside of this Code provide a different definition.

1.4.9 The following words and terms in this Code shall be interpreted in accordance with (a) through (e):
(a) The word "shall" is always mandatory.

COMMENTARY

R1.3.2 The minimum requirements in this Code do not replace sound professional judgment or the licensed design professional's knowledge of the specific factors surrounding a project, its design, the project site, and other specific or unusual circumstances to the project.

CODE

(b) Provisions of this Code are mandatory even if the word "shall" is not used.
(c) Words used in the present tense shall include the future.
(d) The word "and" indicates that all of the connected items, conditions, requirements, or events shall apply.
(e) The word "or" indicates that the connected items, conditions, requirements, or events are alternatives, at least one of which shall be satisfied.

1.4.10 In any case in which one or more provisions of this Code are declared by a court or tribunal to be invalid, that ruling shall not affect the validity of the remaining provisions of this Code, which are severable. The ruling of a court or tribunal shall be effective only in that court's jurisdiction and shall not affect the content or interpretation of this Code in other jurisdictions.

1.5—Authority having jurisdiction
1.5.1 All references in this Code to the authority having jurisdiction shall be understood to refer to the persons who administer and enforce this Code.

1.6—Licensed design professional
1.6.1 All references in this Code to the licensed design professional shall be understood to refer to the persons who possess the knowledge, judgment, and skills to interpret and properly use this Code and are licensed in the jurisdiction where this Code is being used. The licensed design professional is responsible for, and in charge of, the design of the strengthening measures using FRP systems.

1.7—Inspector
1.7.1 All references in this Code to the inspector shall be understood to refer to the persons employed or retained by an approved agency and approved by the authority having jurisdiction as having the competence necessary to observe, test, or otherwise evaluate field installations of FRP systems for compliance with the construction documents.

1.8—Design and construction documents
1.8.1 Documentation of FRP system design shall satisfy 1.8.1.1 or 1.8.1.2.

1.8.1.1 The required performance of the FRP strengthened structural system shall be specified in the construction documents.

1.8.1.2 Complete details of the FRP system shall be depicted on the construction documents.

COMMENTARY

R1.6—Licensed design professional
R1.6.1 The design of FRP systems is often delegated to a licensed design professional with expertise in this work, commonly known as a specialty engineer. Others, for example the engineer of record for a rehabilitation project, are often responsible for the condition assessment that identifies the need for strengthening, the strengthening requirements, and for determining whether an FRP strengthening system is an appropriate option for the application.

R1.8—Design and construction documents
R1.8.1 Both methods are described in ACI SPEC-440.12 and are applicable to all FRP strengthening systems.

R1.8.1.1 This method requires the FRP system performance to be specified by the licensed design professional and design calculation to be submitted as a deferred submittal demonstrating that the required performance is achieved. Information pertaining to required performance typically includes existing material strength, existing and required strengthened capacity, dimensions, and considerations for environmental exposure and fire.

CODE

1.8.2 The construction documents for strengthening using FRP systems shall convey the location, nature, and extent of the work, and the necessary information to perform the work in conformance with the requirements of this Code and the authority having jurisdiction.

1.8.3 Construction documents shall require that materials used for the work conform to this Code and governing regulatory requirements in effect at the time an application for construction of the project is submitted to the authority having jurisdiction.

1.8.4 Calculations pertinent to design shall be submitted with the construction documents if required by the authority having jurisdiction.

1.9—Testing and inspection

1.9.1 Testing and inspection of concrete structures strengthened with FRP systems shall be conducted in accordance with Chapter 11.

1.9.2 Testing and inspection of FRP systems and constituent materials shall be conducted in accordance with Chapter 11 to ensure compliance with Chapter 4.

1.9.3 Testing and inspection of concrete substrates prior to the application of FRP systems shall be conducted in accordance with Chapter 11 to ensure compliance with Chapter 5.

1.10—Approval of special systems of design, construction, or alternative construction materials

1.10.1 Alternate FRP strengthening systems not within the scope of this Code shall be permitted to be used if equivalent strength, serviceability, and durability are demonstrated by testing and analysis, to satisfaction of the authority having jurisdiction.

1.10.2 FRP systems that are approved by the authority having jurisdiction through alternative means and methods clauses in the design basis code shall be permitted.

COMMENTARY

R1.10—Approval of special systems of design, construction, or alternative construction materials

R1.10.1 Results from load tests, large- or full-scale model tests loaded to failure, and other types of physical testing can be used to supplement analytical procedures in the evaluation or design of FRP systems or their use in strengthening existing structures.

CODE

CHAPTER 2—NOTATION AND DEFINITIONS

2.1—Code notation

A_{anc}	=	gross laminate area of the fiber anchor, in.²
A_e	=	cross-sectional area of effectively confined concrete section, in.²
A_f	=	nominal area of FRP reinforcement, in.²
$A_{fanchor}$	=	area of transverse FRP U-wrap for anchorage of flexural FRP reinforcement, in.²
A_{fv}	=	area of FRP shear reinforcement with spacing s_f, in.²
A_g	=	gross area of concrete section, in.²
A_{st}	=	total area of nonprestressed longitudinal steel reinforcement excluding prestressing reinforcement, in.²
a_b	=	smaller cross-sectional dimension for rectangular FRP, in.
b_b	=	larger cross-sectional dimension for rectangular FRP, in.
b_c	=	short side dimension of compression member of prismatic cross section, in.
b_w	=	web width or diameter of circular section, in.
C_E	=	environmental reduction factor
c	=	distance from extreme compression fiber to the neutral axis, in.
D_c	=	diameter of circular cross section, in.
d	=	distance from extreme compression fiber to centroid of tension reinforcement, in.
d_b	=	diameter of round NSM bar, in.
d_{fv}	=	effective depth of FRP shear reinforcement, in.
E_2	=	slope of linear portion of stress-strain model for FRP-confined concrete, psi
E_c	=	modulus of elasticity of concrete, psi
E_f	=	design chord tensile modulus of elasticity of FRP, psi
E_f^*	=	chord tensile modulus of elasticity of FRP, psi
E_s	=	modulus of elasticity of steel, psi
f_c	=	compressive stress in concrete, psi
f_c'	=	specified compressive strength of concrete, psi
f_{cc}'	=	compressive strength of confined concrete, psi
$f_{c,s}$	=	compressive stress in concrete at service condition, psi
f_{fd}	=	design stress of externally bonded FRP reinforcement, psi

COMMENTARY

CHAPTER R2—NOTATION AND DEFINITIONS

R2.1—Commentary notation

A_p	=	area of prestressed reinforcement in tension zone, in.²
A_s	=	area of nonprestressed longitudinal steel tension reinforcement, in.²
b	=	width of compression face of member, in.
C_c	=	resultant force of concrete stress block, lb
d_f	=	effective depth of FRP flexural reinforcement, in.
d_p	=	distance from extreme compression fiber to centroid of prestressed reinforcement, in.
e	=	eccentricity of prestressing steel with respect to centroidal axis of member at support, in.

CODE

f_{fe} = effective stress in the FRP; stress attained at section failure, psi
$f_{f,s}$ = stress in FRP caused by a moment within elastic range of member, psi
f_{fu} = design ultimate tensile strength of FRP, psi
$\overline{f}_{fu}$ = mean tensile strength of a sample of at least 20 FRP test specimens
f_{fu}^* = ultimate tensile strength of the FRP material, psi
f_ℓ = maximum confining pressure due to FRP jacket, psi

$f_{ps,s}$ = stress in prestressed reinforcement at service load, psi
f_{pu} = specified tensile strength of prestressing reinforcement, psi
f_{py} = specified yield strength of prestressing reinforcement, psi

$f_{s,s}$ = stress in nonprestressed steel reinforcement at service loads, psi
f_y = specified yield strength of nonprestressed steel reinforcement, psi
h = overall thickness, height, or depth of member
h_{anc} = minimum embedment depth of fiber anchor in concrete, in.
h_c = long side cross-sectional dimension of rectangular compression member, in.

k_1 = modification factor applied to κ_v to account for concrete strength
k_2 = modification factor applied to κ_v to account for wrapping scheme
L_a = total length of unbonded tendon between anchorages, in.
L_e = active bond length of FRP laminate, in.
ℓ_{df} = development length of FRP system, in.
M_{cr} = cracking moment, in.-lb
M_n = nominal flexural strength at section, in.-lb

M_u = factored moment at a section, in.-lb
N = number of plies of FRP reinforcement

P_n = nominal axial compressive strength of a concrete section, lb
P_u = factored axial force; to be taken as positive for compression and negative for tension, lb

COMMENTARY

f_{ps} = stress in prestressed reinforcement at nominal strength, psi

f_s = stress in nonprestressed steel reinforcement, psi

I_{cr} = moment of inertia of cracked section transformed to concrete, in.^{4}
I_g = moment of inertia of gross section about centroidal axis, in.^{4}
k = ratio of depth of neutral axis to reinforcement depth measured from extreme compression fiber

M_s = service moment at section, in.-lb
M_{snet} = service moment at section beyond decompression, in.-lb

P_e = effective force in prestressing reinforcement (after allowance for all prestress losses), lb

CODE

R_A = ratio of area of fiber in fiber anchor to area of fiber in one leg of shear U-wrap

r_{anc} = length of fiber anchor splay, in.

r_c = radius of edges or corners of a prismatic cross section confined with FRP, in.

s_{anc} = center-to-center spacing of fiber anchors, in.

s_f = center-to-center spacing of FRP strips, in.

T_g = glass transition temperature, °F

t_f = nominal thickness of one ply of FRP reinforcement, in.

V_c = nominal shear strength provided by concrete, lb

V_f = nominal shear strength provided by FRP stirrups, lb

V_n = nominal shear strength, lb

V_s = nominal shear strength provided by steel shear reinforcement, lb

V_u = factored shear force at section, lb

w_f = width of FRP reinforcing plies, in.

α = angle of application of primary FRP reinforcement direction relative to longitudinal axis of member, degrees

α_{anc} = angle over which fiber anchor is splayed over externally bonded FRP, degrees

β_{anc} = embedment angle of fiber anchor, degrees

ε_{bi} = strain in concrete substrate at time of FRP installation (tension is positive), in./in.

ε_c = strain in concrete, in./in.

ε_c' = compressive strain of unconfined concrete corresponding to f_c', in./in.

ε_{ccu} = ultimate axial compressive strain of confined concrete corresponding to $0.85f_{cc}'$ in a lightly confined member (member confined to restore its concrete design compressive strength), or ultimate axial compressive strain of confined concrete corresponding to failure in a heavily confined member, in./in.

ε_{ct} = concrete tensile strain at level of tensile force resultant in post-tensioned flexural members, in./in.

COMMENTARY

r = radius of gyration of cross section, in.

T_f = resultant force of FRP system, lb

T_s = resultant force of internal reinforcing steel, lb

T_y = resultant force of internal reinforcing steel assuming yield of steel, lb

y_b = distance from centroidal axis of gross section, neglecting reinforcement, to extreme bottom fiber, in./in.

α_1 = multiplier on f_c' to determine intensity of an equivalent rectangular stress distribution for concrete

β_1 = ratio of depth of equivalent rectangular stress block to depth of the neutral axis

ε_{cu} = ultimate axial strain of unconfined concrete corresponding to $0.85f_c'$ or maximum usable strain of unconfined concrete, in./in., which can

CODE

ε_{fd} = debonding strain of externally bonded FRP reinforcement, in./in.
ε_{fe} = effective strain in FRP reinforcement attained at failure, in./in.
ε_{fu} = design rupture strain of FRP reinforcement, in./in.
ε_{fu}^* = rupture strain of FRP reinforcement, in./in.
ε_{pe} = effective strain in prestressing steel after losses, in./in.

ε_{ps} = strain in prestressed reinforcement at nominal strength, in./in.

ε_{sy} = strain corresponding to yield strength of nonprestressed steel reinforcement, in./in.
ε_t = net tensile strain in extreme tension steel reinforcement at nominal strength, in./in.
ε_t' = transition strain in stress-strain curve of FRP-confined concrete, in./in.
η = parameter that combines the effects of member continuity and applied load pattern for producing maximum factored moment at the critical section under consideration
ϕ = strength reduction factor
κ_a = shape factor for FRP reinforcement in determination of f_{cc}' (based on geometry of cross section)
κ_b = shape factor for FRP reinforcement in determination of ε_{ccu} (based on geometry of cross section)
κ_v = bond-dependent coefficient for shear
ρ_g = ratio of area of longitudinal steel reinforcement to cross-sectional area of a compression member (A_s/bh)
σ = standard deviation
τ_b = average bond strength for near-surface-mounted FRP bars, psi
ψ_f = FRP strength reduction factor

2.2—Definitions

COMMENTARY

occur at $f_c = 0.85f_c'$ or $\varepsilon_c = 0.003$, depending on the obtained stress-strain curve

ε_{pi} = initial strain in prestressed steel reinforcement, in./in.
ε_{pnet} = net strain in flexural prestressing steel at limit state after prestress force is discounted (excluding strains due to effective prestress force after losses), in./in.
$\varepsilon_{pnet,s}$ = net strain in prestressing steel beyond decompression at service, in./in.

$\varepsilon_{ps,s}$ = strain in prestressing steel at service load, in./in.
ε_s = strain in nonprestressed steel reinforcement, in./in.

R2.2—Definitions

For consistent application of this Code, it is necessary that terms be defined where they have particular meaning in this Code. The definitions given are for use in application of this Code only and do not always correspond to ordinary usage. A glossary of most-used terms relating to cement manufacturing, concrete design and construction,

CODE

bond critical—application of strengthening system that relies on load transfer from the substrate to the system material achieved through bond stresses at the interface.

carbon fiber—fiber produced by heating organic precursor materials containing a substantial amount of carbon, such as rayon, polyacrylonitrile, or pitch in an inert environment.

carbon fiber-reinforced polymer—composite material comprising a polymer matrix reinforced with carbon fiber fabric, mat, or strands.

construction documents—written and graphic documents and specifications prepared or assembled that describe the location, design, materials, and physical characteristics of the elements of a project necessary for obtaining a building permit and construction of the project.

contact critical—application of strengthening system that relies on load transfer from the substrate to the system material achieved through contact or bearing at the interface.

creep rupture—breakage of a material under sustained loading at stresses less than the tensile strength.

cure—process by which the components of a thermosetting resin react to produce specified properties.

design basis code—legally adopted code requirements under which the assessment and strengthening measures are designed and constructed.

externally bonded FRP system—wet layup or precured FRP strengthening system applied to the surface of a concrete member.

fabric—two-dimensional network of woven, nonwoven, knitted, or stitched fibers; yarns; or tows.

fiber anchor—bundle of fibers having a specified length and fiber content and impregnated with resin; a portion of the anchor is embedded into the concrete substrate and the remaining portion is splayed over externally bonded fabric.

fiber content—amount of fiber present in a composite, expressed as a percentage volume fraction or mass fraction of the composite.

fiber volume fraction—ratio of the volume of fibers to the volume of the composite containing the fibers.

glass fiber—filament drawn from an inorganic fusion typically comprising silica-based material that has cooled without crystallizing.

glass fiber-reinforced polymer—composite material comprising a polymer matrix reinforced with glass fiber fabric, mat, or strands.

glass transition temperature—representative temperature of the temperature range over which an amorphous material (such as glass or a polymer) changes from (or to) a brittle, vitreous state to (or from) a plastic state.

impregnate—to saturate fibers with resin or binder.

COMMENTARY

and research in concrete is contained in "ACI Concrete Terminology," available on the ACI website. Additional definitions relevant to FRP strengthening systems can be found in ACI PRC-440.2.

FRP systems are conventionally supplied as systems to ensure the compatibility of all constituent components.

CODE

intumescent coating—covering that swells, increasing volume and decreasing density, when exposed to fire imparting a degree of passive fire protection.

jacket—FRP system that fully encloses a concrete column member cross section and has a lap splice adequate to develop the capacity of the FRP.

laminate—multiple plies or layers of fiber reinforcement molded together.

near-surface-mounted FRP system—precured FRP strengthening system installed into grooves cut into concrete surface.

precured FRP system—collective term for all components of an FRP strengthening application that may include surface primer, adhesive, precured FRP, and protective overcoats, as applicable.

putty—thickened polymer-based resin used to prepare the concrete substrate.

saturating resins—polymer-based resin used to impregnate the reinforcing fibers, bond them in place, and transfer load between fibers.

vinyl ester resin—thermosetting reaction product of epoxy resin with a polymerizable unsaturated acid (usually methacrylic acid) that is then diluted with a reactive monomer (usually styrene).

wet layup—manufacturing process where dry fabric reinforcement is impregnated on site with a saturating resin matrix and then cured in place.

wet layup fiber-reinforced polymer system—collective term for all components of an FRP strengthening application that may include surface primer, putty, saturant, fiber fabric, mat or strands, and protective overcoats, as applicable.

witness panel—small mockup manufactured under conditions representative of field application, to confirm that prescribed procedures and materials will yield specified mechanical and physical properties.

COMMENTARY

FRP systems are conventionally supplied as systems to ensure the compatibility of all constituent components.

FRP systems are conventionally supplied as systems to ensure the compatibility of all constituent components.

FRP systems are conventionally supplied as systems to ensure the compatibility of all constituent components.

CODE

CHAPTER 3—REFERENCED STANDARDS

3.1—American Concrete Institute

ACI CODE-216.1-14—Code Requirements for Determining Fire Resistance of Concrete and Masonry Construction Assemblies

ACI CODE-318-19(22)—Building Code Requirements for Structural Concrete and Commentary

ACI SPEC-440.6-08(22)—Specification for Carbon Fiber-Reinforced Polymer Bar Material for Concrete Reinforcement

ACI SPEC-440.8-13(23)—Specification for Carbon and Glass Fiber-Reinforced Polymer Materials Made by Wet Layup for External Strengthening

ACI SPEC-440.12-22—Specification for Strengthening of Concrete Structures with Externally Bonded Fiber-Reinforced Polymer (FRP) Materials using the Wet Layup Method

ACI CODE-562-21—Assessment, Repair, and Rehabilitation of Existing Concrete Structures—Code and Commentary

3.2—American Society of Civil Engineers

ASCE/SEI 7-22—Minimum Design Loads and Associated Criteria for Buildings and Other Structures

3.3—ASTM International

ASTM C882/C882M-20—Standard Test Method for Bond Strength of Epoxy-Resin Systems Used With Concrete By Slant Shear

ASTM C1583/C1583M-13—Standard Test Method for Tensile Strength of Concrete Surfaces and the Bond Strength or Tensile Strength of Concrete Repair and Overlay Materials by Direct Tension (Pull-off Method)

ASTM D570-22—Standard Test Method for Water Absorption of Plastics

ASTM D638-22—Standard Test Method for Tensile Properties of Plastics

ASTM D695-15—Standard Test Method for Compressive Properties of Rigid Plastics

ASTM D3039/3039M-17—Standard Test Method for Tensile Properties of Polymer Matrix Composite Materials

ASTM D3418/D3418M-03—Standard Test Method for Transition Temperatures and Enthalpies of Fusion and Crystallization of Polymers by Differential Scanning Calorimetry

ASTM D4258-05(2017)—Standard Practice for Surface Cleaning Concrete for Coating

ASTM D4259-18—Standard Practice for Preparation of Concrete by Abrasion Prior to Coating Application

ASTM D7565/D7565M-10(2017)—Standard Test Method for Determining Tensile Properties of Fiber Reinforced Polymer Matrix Composites Used for Strengthening of Civil Structures

ASTM D7682-17—Standard Test Method for Replication and Measurement of Concrete Surface Profiles Using Replica Putty

COMMENTARY

CODE

ASTM D7957/D7957M-22—Standard Specification for Solid Round Glass Fiber Reinforced Polymer Bars for Concrete Reinforcement

ASTM E119-22—Standard Test Methods for Fire Tests of Building Construction and Materials

3.4—Underwriters Laboratories

ANSI/UL 263-11—Fire Tests of Building Construction and Materials

COMMENTARY

CODE

CHAPTER 4—FRP SYSTEM REQUIREMENTS

4.1—General

4.1.1 FRP systems shall conform to the requirements of 4.2, 4.3, or 4.4 and be reported based on gross laminate area or nominal bar area.

4.1.2 Gross laminate area shall be taken to be equal to laminate thickness t_f multiplied by the laminate width.

4.2—Wet layup FRP systems

4.2.1 Wet layup FRP systems shall conform to the requirements of ACI SPEC-440.8.

4.3—Precured carbon FRP systems

4.3.1 For precured carbon FRP plates, ultimate tensile strength and mean chord tensile modulus shall be determined in accordance with ASTM D7565/D7565M or ASTM D3039/D3039M.

4.3.2 The ultimate tensile strength of precured carbon FRP, determined as the mean tensile strength of a sample of at least 20 specimens minus three times the standard deviation ($f_{fu}^* = \overline{f}_{fu} - 3\sigma$), shall not be less than 170 ksi.

COMMENTARY

CHAPTER R4—FRP SYSTEM REQUIREMENTS

R4.1—General

R4.1.1 ACI PRC-440.2 defines the ultimate tensile strength of an FRP strengthening system (f_{fu}^*) as the mean tensile strength of a sample of at least 20 specimens minus three times the standard deviation ($f_{fu}^* = \overline{f}_{fu} - 3\sigma$). The chord tensile modulus (E_f^*) is defined as the mean chord tensile modulus of a sample of at least 20 test specimens. From these, rupture strain is determined: $\varepsilon_{fu}^* = f_{fu}^*/E_f^*$.

R4.2—Wet layup FRP systems

R4.2.1 ACI SPEC-440.8 is applicable to carbon and glass fiber systems applied using the wet layup method with an epoxy resin. ACI SPEC-440.8 reports mechanical properties determined in accordance with ASTM D7565/D7565M in terms of tensile force and chord tensile modulus per unit width of material. To obtain the required strength and modulus in terms of gross laminate area, the reported tensile force and chord tensile modulus per unit width values are divided by the nominal laminate thickness. ASTM D7565/D7565 prescribes tensile testing in accordance with ASTM D3039/D3039M. Results from ASTM D3039/D3039M testing, which uses gross laminate area, can be reported directly.

Based on the requirements of ACI SPEC-440.8, minimum ultimate tensile strength and mean chord tensile modulus requirements for wet layup FRP systems are 55 ksi and 2500 ksi, respectively.

R4.3—Precured carbon FRP systems

R4.3.1 Precured carbon FRP plates are used for externally bonded and, when cut into strips, near-surface-mounted (NSM) applications. Mechanical properties determined in accordance with ASTM D7565/D7565M are reported in terms of tensile force and chord tensile modulus per unit width of material. To obtain the required strength and modulus in terms of gross laminate area, the reported tensile force and chord tensile modulus per unit width values are divided by the nominal FRP laminate thickness. ASTM D7565/D7565 prescribes tensile testing in accordance with ASTM D3039/D3039M. Results from ASTM D3039/D3039M testing, which uses gross laminate area, can be reported directly.

CODE

4.3.3 The mean chord tensile modulus of precured carbon FRP (E_f^*), determined from a sample of at least 20 test specimens shall not be less than 18,000 ksi.

4.3.4 The rupture strain shall be determined from Eq. (4.3.4).

$$\varepsilon_{fu}^* = f_{fu}^*/E_f^* \qquad (4.3.4)$$

4.3.5 Adhesive used to secure precured carbon FRP and FRP NSM reinforcement to concrete substrate shall conform to the requirements given in Table 4.3.5 when cured at 73 ± 3°F.

Table 4.3.5—Physical and mechanical property requirements of adhesive

Property	ASTM standard	Required value
Minimum bond strength, 14 days	C882/C882M	1000 psi
Maximum absorption, 24 hours	D570	1%
Minimum glass transition temperature, 7 days	D3418/D3418M	140°F
Minimum compressive yield strength, 7 days	D695	8000 psi
Minimum compressive modulus, 7 days	D695	150,000 psi
Minimum tensile strength, 7 days	D638	3600 psi
Minimum elongation at break, 7 days	D638	1%

4.3.5.1 Specimens used for determination of minimum glass transition temperature, minimum compressive yield strength, minimum compressive modulus, minimum tension strength, and minimum elongation at break shall be permitted to be post-cured for up to 72 hours at 140°F. In such cases, test results shall be reported with and without post-cure.

4.4—FRP near-surface-mounted bars

4.4.1 Glass FRP NSM bars shall be No. 4 or smaller and conform to ASTM D7957/D7957M.

Table R4.4.1—Nominal geometry and minimum mechanical properties of glass FRP NSM bars (ASTM D7957/D7957M)

Bar designation	No. 2	No. 3	No. 4
Nominal diameter, in.	0.250	0.375	0.500
Nominal area, in.²	0.049	0.11	0.20
Minimum ultimate tension force, kip	6.1	13.2	21.6
Minimum ultimate tensile strength, f_{fu}^*, ksi	124	120	108
Minimum mean chord tensile modulus, E_f^*, ksi	6500		
Rupture strain, ε_f^*	> 0.011		

COMMENTARY

R4.3.5.1 The permitted post cure is intended to better reflect physical and mechanical properties of the cured FRP system.

R4.4—FRP near-surface-mounted bars

R4.4.1 ASTM D7957/D7957M is applicable to only bars having glass fibers in a vinyl ester matrix resin. ASTM D7957/D7957M prescribes a minimum bar rupture force. The implied strength is obtained by dividing the force by the nominal area of the bar. Table R4.4.1 gives minimum ultimate tensile strength and mean chord tensile modulus requirements of ASTM D7957/D7957M. Note that the ultimate tensile strength decreases with increased bar diameter.

CODE

4.4.2 Carbon FRP NSM bars shall be No. 4 or smaller and conform to the material properties set forth in ACI SPEC-440.6.

COMMENTARY

R4.4.2 ACI SPEC-440.6 is applicable only to bars having carbon fibers in either a vinyl ester or an epoxy matrix resin. ACI SPEC-440.6 prescribes a minimum bar rupture force. The strength is obtained by dividing the force by the nominal area of the bar. Table R4.4.2 gives minimum ultimate tensile strength and mean chord tensile modulus requirements of ACI SPEC-440.6. Note that the ultimate tensile strength decreases with increased bar diameter.

Table R4.4.2—Nominal geometry and minimum mechanical properties of carbon FRP NSM bars (ACI SPEC-440.6)

Bar designation	No. 2	No. 3	No. 4
Nominal diameter, in.	0.250	0.375	0.500
Nominal area, in.^{2}	0.05	0.11	0.20
Minimum ultimate tensile strength, f_{fu}^*, ksi	210	190	170
Minimum mean chord tensile modulus, E_f^*, ksi	18,000		
Rupture strain, ε_f^*	> 0.005		

4.4.3 Adhesive used to embed NSM FRP into concrete grooves shall conform to the requirements of 4.3.5.

CODE	COMMENTARY
CHAPTER 5—CONCRETE SUBSTRATE REQUIREMENTS	**CHAPTER R5—CONCRETE SUBSTRATE REQUIREMENTS**

CODE

5.1—General

5.1.1 Concrete surface preparation shall be in accordance with the construction documents and shall meet the minimum requirements of 5.2, 5.3, or 5.4.

5.1.2 The compressive strength of the concrete substrate shall not be less than 2500 psi.

5.1.3 The tensile strength of the substrate concrete, determined in accordance with ASTM C1583/C1583M shall not be less than 200 psi.

5.1.4 Substrates that exhibit deterioration shall be evaluated. Remediation of substrate to address existing damage and mitigate future deterioration shall be implemented prior to application of FRP systems.

5.2—Bond critical externally bonded FRP systems

5.2.1 For bond critical externally bonded FRP systems, the concrete substrate and substrate preparation shall conform to requirements of 5.2.1.1 through 5.2.1.4.

5.2.1.1 Methods of surface preparation shall be in accordance with ASTM D4259. The minimum surface profile, determined in accordance with ASTM D7682 Method A or by direct visual comparison, shall be CSP 3.

5.2.1.2 The substrate shall be cleaned in accordance with ASTM D4258.

5.2.1.3 Localized out-of-plane variations, including form lines, shall not exceed 1/32 in.

5.2.1.4 Where FRP systems wrap around corners, the corners shall be rounded to a radius not less than 0.5 in.

COMMENTARY

R5.1—General

R5.1.1 The performance of concrete members strengthened with FRP systems is highly dependent on proper preparation and profiling of the concrete substrate. Procedures for installing FRP systems have been developed by the system manufacturers and often differ between systems; these often include surface preparation requirements. Deviations from the procedures developed by the FRP system manufacturer should not be permitted without consulting with the licensed design professional and the FRP system manufacturer.

R5.1.4 Corrosion of steel reinforcement, effects of freezing-and-thawing cycles, chemical reactions such as ASR, cracks exceeding 0.01 in. width (ACI PRC-440.2), and others, can cause substrate deterioration. Prior to specifying an FRP system, the licensed design professional should investigate the severity and extent of existing deterioration and potential for future deterioration. Based on the investigation findings, the licensed design professional should develop a scope for remedial work to address existing and to mitigate future substrate deterioration. The investigation may reveal that the use of an FRP system is not suitable due to the extent and type of deterioration. ACI PRC-546 provides recommendations for the selection and application of materials and methods for repairing concrete substrates.

R5.2—Bond critical externally bonded FRP systems

R5.2.1.1 ASTM D7682 uses putty impressions to capture the surface profile of prepared concrete. Method A describes a direct visual comparison of the impression with the surface profile chips accompanying ICRI 310.2R.

CODE

5.3—Contact critical externally bonded FRP systems

5.3.1 For externally bonded contact-critical FRP systems, the concrete substrate and substrate preparation shall conform to 5.2.1.2 through 5.2.1.4.

5.4—NSM FRP systems

5.4.1 The interior surfaces of NSM grooves shall be prepared in accordance with the system manufacturer's requirements.

5.4.2 The geometry and spacing of NSM grooves shall comply with 7.7.2.

COMMENTARY

R5.4—NSM FRP systems

R5.4.1 The interior surface of the groove should typically be cleaned in accordance with ASTM D4258.

CODE

CHAPTER 6—GENERAL DESIGN REQUIREMENTS

6.1—General
6.1.1 This chapter shall apply to the design requirements for strengthening existing concrete structures using FRP systems.

6.2—Basis of design
6.2.1 Design shall be based on the dimensions, amount, distribution, and locations of internal steel reinforcement, material properties, and condition of the existing concrete member to be strengthened.

6.2.2 Structural elements strengthened with FRP systems shall have design strengths at all sections at least equal to the required strengths calculated using the applied factored loads and internal forces in such combinations as required by the design basis code.

6.2.3 Structural elements supporting FRP-strengthened members or connections shall have design strengths at all sections at least equal to the required strengths calculated using the applied factored loads and internal forces in such combinations as required by the design basis code.

6.2.4 Bond critical FRP strengthening systems shall be designed to resist tensile forces while maintaining strain compatibility between the FRP and concrete substrate.

6.2.5 Design strains in the FRP system shall consider the state of strain of the substrate concrete when the FRP system is installed.

6.2.6 FRP strengthening systems shall not be designed to resist compressive forces.

6.3—Load factors and combinations
6.3.1 Load factors and combinations for FRP strengthening systems shall be in accordance with ACI CODE-562-21 Chapter 5 requirements including those for structures rehabilitated with external reinforcing systems.

6.3.2 The strength of the concrete element without FRP strengthening shall satisfy ACI CODE-562-21 Section 5.5.2.

COMMENTARY

CHAPTER R6—GENERAL DESIGN REQUIREMENTS

R6.2—Basis of design
R6.2.1 ACI CODE-562 provides requirements for evaluation and design of members to be strengthened.

R6.2.2 Applications involving increasing the design loads on a structure will require structural analysis to determine adequate strength at various critical sections of existing elements. For example, members strengthened for flexure need to also have adequate shear capacity to resist the imposed loads.

R6.2.3 Members supporting strengthened members also need to possess adequate strength. For example, columns and foundations need to be capable of sustaining the loads from strengthened beam elements.

R6.2.4 Bond critical FRP strengthening systems contribute to the load carrying capacity of a concrete member as additional tensile reinforcement. Tensile strains induced in the FRP induce stress in the FRP, which result in a resistive force. Strain compatibility and force equilibrium for various applications are addressed in Chapters 7 through 9 of this Code.

R6.2.5 FRP systems are only affected by loads imposed after the FRP system is installed and cured. For example, the FRP system does not resist the effects of structure self-weight present while the FRP system is installed.

R6.2.6 It is acceptable for FRP strengthening systems to experience compressive forces. However, any resistance to compressive force that the FRP may provide is neglected in the provisions of this Code.

R6.3—Load factors and combinations

R6.3.2 This requirement is intended to minimize the risk of failure of the strengthened structural member in the case where, during normal operating conditions, the external reinforcement is damaged. If such damage is not detected immediately, the ability of the structure (or component) to resist full design loads may be compromised until the

CODE

6.3.3 If a fire-resistance rating is required by the design basis code, FRP-strengthened concrete members shall satisfy ACI CODE-562-21 Section 5.5.3.

6.4—Design material properties

6.4.1 Design tensile properties of FRP strengthening systems shall be determined from Eq. (6.4.1a) through (6.4.1c) using the mechanical properties defined in Chapter 4 and the environmental reduction factor, CE, defined in 6.4.2.

$$f_{fu} = C_E f_{fu}^* \quad (6.4.1a)$$

$$\varepsilon_{fu} = C_E \varepsilon_{fu}^* \quad (6.4.1b)$$

$$E_f = f_{fu}/\varepsilon_{fu} \quad (6.4.1c)$$

6.4.1.1 The requirements of 6.4.1 shall be applied regardless of the use of protective coatings on the FRP system.

6.4.2 Environmental reduction factors shall be determined according to Table 6.4.2.

Table 6.4.2—Environmental reduction factors for various exposure conditions and fiber types

Exposure condition	Fiber type	Environmental reduction factor C_E
Interior exposure	Carbon	0.95
	Glass	0.75
Exterior exposure	Carbon	0.85
	Glass	0.65
Aggressive environment	Carbon	0.85
	Glass	0.50

COMMENTARY

damage is identified and addressed. Appendix A describes additional load combinations in ACI CODE-562 that apply to externally bonded FRP strengthening systems.

R6.3.3 This requirement is intended to ensure that the repaired element will maintain sufficient strength, accounting for its probable reduced material properties due to elevated temperatures, during a fire event. Appendix A describes additional load combinations in ACI CODE-562 that apply to externally bonded FRP strengthening systems.

R6.4—Design material properties

R6.4.1 Tensile properties reported by manufacturers do not consider long-term exposure to environmental conditions. Because long-term exposure to various environments can degrade the tensile strength, rupture strain, and the creep and fatigue performance of FRP strengthening systems, the material properties used in design equations need to be reduced by an environmental reduction factor, CE (ACI PRC-440.2). Environmental conditions affect both strength and strain but have been shown to have a limited effect on the tensile modulus of FRP strengthening systems (Cromwell et al. 2011).

R6.4.2 Interior exposure applies to applications in temperature and humidity-controlled spaces free of aggressive environments. Interior exposure is typically an indoor application of the FRP strengthening system that will be protected from the elements such as interior occupied spaces.

Exterior exposure applies to applications in outdoor environments free of aggressive environments. Exterior exposure is typically outdoor exposure associated with applications such as bridges, piers/wharfs above the high-water line, and parking garages.

Aggressive environment applies to applications involving exposure to any one or more of the following conditions:
(a) Temperatures exceeding 120°F for a duration exceeding 24 consecutive hours
(b) Temperatures below –20°F for a duration exceeding 24 consecutive hours
(c) Temperatures exceeding 100°F combined with humidity levels exceeding 95% relative humidity for a duration exceeding 24 consecutive hours
(d) Continuous immersion in water
(e) Continuous contact with soil

Aggressive environments are those associated with conditions known to cause greater long-term degradation of material properties or bond properties. Exposure to chemical environments requires additional consideration beyond the scope of this Code, including the specification of additional protective coatings.

CODE

6.5—Maximum sustained loads

6.5.1 The sum of sustained stress plus cyclic stress in the FRP strengthening system shall not exceed $0.55f_{fu}$ for carbon FRP and $0.20f_{fu}$ for glass FRP.

6.6—Maximum service temperature

6.6.1 FRP strengthening systems shall not be used when in-service substrate temperatures are expected to exceed the glass transition temperature of the FRP strengthening system minus 27°F ($T_g - 27$°F).

6.7—Durability requirements

6.7.1 Means to allow moisture to escape from the existing structure shall be provided when the FRP strengthening system creates an impermeable layer and moisture vapor transmission is expected.

6.7.2 Carbon FRP strengthening systems shall be designed to be electrically isolated from any metallic components of the structure, including reinforcing steel.

6.7.3 The need for protective coatings shall be based on the anticipated environmental exposure conditions and their compatibility with the FRP strengthening system.

COMMENTARY

R6.5—Maximum sustained loads

R6.5.1 Carbon fiber FRP strengthening systems are generally used in applications where resistance to fatigue and sustained load is required.

R6.6—Maximum service temperature

R6.6.1 This requirement is intended for elevated service temperatures such as those found in hot climatic regions or certain industrial environments. The physical and mechanical properties of the resin components of the FRP strengthening system are influenced by temperature and moisture and degrade at temperatures close to or above their glass transition temperature T_g (ACI PRC-440.2). The reported T_g is taken as the lowest value of T_g of the components of the system.

6.7—Durability requirements

R6.7.1 FRP strengthening systems may create a moisture barrier. Moisture and moisture vapor escaping the existing structure that is trapped by the FRP can cause debonding of the FRP strengthening system; thus, adequate means of moisture escape is needed. This is often achieved by leaving gaps of exposed concrete substrate between strips of the FRP strengthening system. Weep channels or weep holes in the installed system can also be employed to allow moisture egress.

R6.7.2 Galvanic corrosion can initiate when carbon fiber is in electrical contact with metals such as steel. In most strengthening applications, the saturant or epoxy adhesive used to apply the FRP material to the concrete substrate is adequate to isolate the metal. Special care should be taken to ensure electrical isolation of carbon FRP and metallic inserts in the substrate surface. In such cases, an additional layer of glass FRP is typically adequate.

R6.7.3 Protective coatings can be used for aesthetic reasons or to protect the FRP system from UV degradation, abrasion, accidental damage, vandalism, reactions to chemical exposure, and other conditions.

CODE

CHAPTER 7—DESIGN AND DETAILING FOR FLEXURAL STRENGTHENING

7.1—General

7.1.1 This chapter shall apply to the design and detailing of flexural strengthening of reinforced concrete and prestressed concrete members when permitted by the provisions of this Code.

7.1.2 Flexural strengthening applications shall be limited to externally bonded and near-surface-mounted (NSM) FRP systems having primary fibers oriented longitudinally in the direction normal to the moment.

7.1.3 Unless anchorage of the flexural strengthening FRP system is provided, FRP reinforcement shall not be used to strengthen concave soffits where the extent of the curved portion exceeds a length of 40 in. with a rise of 0.2 in.

7.2—Design strength

7.2.1 The design strength at all sections shall satisfy Eq. (7.2.1).

$$\phi M_n \geq M_u \qquad (7.2.1)$$

7.2.2 The strength reduction factor ϕ shall be determined in accordance with 7.2.2.1 or 7.2.2.2.

7.2.2.1 For concrete members with nonprestressed steel reinforcement, the strength reduction factor ϕ shall be calculated by Eq. (7.2.2.1).

$$\phi = \begin{cases} 0.90 & \text{for } \varepsilon_t \geq \varepsilon_{sy} + 0.003 \\ 0.65 + \dfrac{0.25(\varepsilon_t - \varepsilon_{sy})}{0.003} & \text{for } \varepsilon_{sy} < \varepsilon_t < \varepsilon_{sy} + 0.003 \\ 0.65 & \text{for } \varepsilon_t \leq \varepsilon_{sy} \end{cases} \qquad (7.2.2.1)$$

7.2.2.2 For prestressed concrete members having bonded prestressed steel reinforcement with an ultimate strength, fpu, of 250 or 270 ksi, the strength reduction factor ϕ shall be calculated by Eq. (7.2.2.2).

$$\phi = \begin{cases} 0.90 & \text{for } \varepsilon_{ps} \geq 0.013 \\ 0.65 + \dfrac{0.25(\varepsilon_{ps} - 0.010)}{0.013 - 0.010} & \text{for } 0.010 < \varepsilon_{ps} < 0.013 \\ 0.65 & \text{for } \varepsilon_{ps} \leq 0.010 \end{cases} \qquad (7.2.2.2)$$

7.2.2.3 For prestressed concrete members having unbonded prestressed steel reinforcement, the strength reduction factor ϕ shall be calculated by Eq. (7.2.2.3).

COMMENTARY

CHAPTER R7—DESIGN AND DETAILING FOR FLEXURAL STRENGTHENING

R7.1—General

R7.1.1 This chapter is applicable to concrete members such as beams, slabs, and walls. Additional strength provided by FRP systems is a function of many factors, including member geometry, strengthening scheme, and existing concrete strength.

R7.1.3 FRP anchors or U-wrap anchorage can be used to delay debonding on concave soffits (Eshwar et al. 2005). The licensed design professional should determine the required anchorage for concave soffits considering the design strain of FRP and anchorage performance data provided by the FRP manufacturer.

R7.2—Design strength

R7.2.1 When varying amounts of tension reinforcement exist along the member span or when draped or harped prestressing steel is present, strength requirements need to be verified at multiple sections along the span.

CODE

$$\phi = \begin{cases} 0.90 & \text{for } \varepsilon_{ct} \geq 0.005 \\ 0.65 + \dfrac{0.25(\varepsilon_{ct} - 0.002)}{0.003} & \text{for } 0.002 < \varepsilon_{ct} < 0.005 \\ 0.65 & \text{for } \varepsilon_{ct} \leq 0.002 \end{cases} \quad (7.2.2.3)$$

7.2.3 The member to be strengthened for flexure shall be capable of resisting the lesser of V_u and the shear forces associated with the increased flexural strength.

7.3—Design requirements

7.3.1 FRP strengthening for flexure shall be designated as being bond critical.

7.3.2 The design of FRP strengthening for flexure shall consider failure modes (a) through (f):
  (a) Crushing of the concrete in compression prior to yielding of reinforcing steel
  (b) Yielding of steel in tension followed by FRP rupture
  (c) Yielding of steel in tension followed by concrete crushing
  (d) Rupture of prestressing steel
  (e) Delamination of the concrete cover
  (f) Debonding of FRP from concrete substrate

7.3.3 Strains in the concrete, nonprestressed steel reinforcement, bonded prestressed steel reinforcement, and FRP reinforcement shall be assumed to be proportional to their distance from the neutral axis of the member.

7.3.3.1 Relative slip between FRP reinforcement and concrete shall be neglected.

7.3.3.2 Shear deformation within the adhesive layer shall be neglected.

7.3.4 The relationship between concrete compressive stress and strain shall be represented by a rectangular, trapezoidal, parabolic, or other shape that results in prediction of strength in substantial agreement with results of comprehensive tests.

COMMENTARY

R7.2.3 The potential for shear failure should be considered by comparing the design shear strength of the section to the required shear strength. If additional shear strength is required, FRP reinforcement designed in accordance with Chapter 8 can be provided.

R7.3—Design requirements

R7.3.2 The flexural strength of a section depends on the controlling failure mode. Cover delamination or FRP debonding can occur if the force in the FRP cannot be transferred through bond to the substrate. Such behavior is generally referred to as debonding, regardless of where the failure plane propagates within the FRP-adhesive-substrate region.

R7.3.3.1 FRP systems that are anticipated to experience post-installation temperature fluctuations greater than ±68°F may need to be evaluated for additional bond stresses induced by the difference in the coefficients of thermal expansion of the FRP and concrete substrate (Lopez et al. 2003; Grace and Singh 2005; Cromwell et al. 2011).

R7.3.3.2 With substrate surface preparation provided in accordance with 5.1 and installation in accordance with ACI SPEC-440.12, the saturant or adhesive layer will be sufficiently thin that the effects of shear deformations are negligible in the context of the design methodology used.

R7.3.4 For the concrete crushing mode of failure, the equivalent rectangular compressive stress distribution as described in ACI CODE-318-19 Sections 22.2.2.4.1 through 22.2.2.4.3 may be used. If FRP rupture, cover delamination, or FRP debonding occur; nonlinear stress distribution in the concrete or a more accurate stress block appropriate for the strain level reached in the concrete at the ultimate-limit state may be used.

When using an equivalent rectangular stress block, internal force equilibrium for a singly reinforced rectangular section is calculated by Eq. (R7.3.4).

CODE

7.3.4.1 Maximum usable compressive strain in the concrete is 0.003.

7.3.4.2 Tensile strength of concrete shall be neglected in flexural strength calculations.

7.3.5 Nonprestressed steel reinforcement shall be assumed to have an elastic-perfectly-plastic stress-strain relationship.

7.3.6 The stress in prestressed steel reinforcement shall be calculated using the material properties of the steel.

7.3.7 The stress in unbonded prestressed steel reinforcement shall be calculated from the effective strain by Eq. (7.3.7).

$$f_{ps} = E_s \varepsilon_{ps} \leq 0.95 f_{py} \quad (7.3.7)$$

7.3.8 The effective stress in the FRP shall be calculated from the effective strain by Eq. (7.3.8).

$$f_{fe} = E_f \varepsilon_{fe} \quad (7.3.8)$$

7.3.8.1 For externally bonded FRP reinforcement, the effective strain ε_{fe} shall not exceed the strain at which debonding occurs, calculated by Eq. (7.3.8.2).

$$\varepsilon_{fd} = 0.083 \sqrt{\frac{f'_c}{N E_f t_f}} \leq 0.9 \varepsilon_{fu} \quad (7.3.8.2)$$

7.3.8.2 For NSM FRP reinforcement, the effective strain ε_{fe} shall not exceed the strain at which debonding occurs calculated by Eq. (7.3.8.3).

$$\varepsilon_{fd} = 0.7 \varepsilon_{fu} \quad (7.3.8.3)$$

COMMENTARY

$$\alpha_1 f'_c \beta_1 bc = A_s f_s + A_p f_{ps} + A_f f_{fe} \quad (R7.3.4)$$

The terms α_1 and β_1 in Eq. (R7.3.4) are parameters defining a rectangular stress block (ACI PRC-440.2).

R7.3.5 The stress in the nonprestressed steel is calculated by Eq. (R7.3.5).

$$f_s = E_s \varepsilon_s \leq f_y \quad (R7.3.5)$$

R7.3.6 For a typical seven-wire low relaxation prestressing strand, the stress-strain curve can be approximated by Eq. (R7.3.6a) or (R7.3.6b) (PCI MNL-120).

For Grade 250 ksi steel:

$$f_{ps} = \begin{cases} 28{,}500 \varepsilon_{ps} & \text{for } \varepsilon_{ps} \leq 0.0076 \\ 250 - \dfrac{0.04}{\varepsilon_{ps} - 0.0064} & \text{for } \varepsilon_{ps} > 0.0076 \end{cases} \quad (R7.3.6a)$$

For Grade 270 steel:

$$f_{ps} = \begin{cases} 28{,}500 \varepsilon_{ps} & \text{for } \varepsilon_{ps} \leq 0.0086 \\ 270 - \dfrac{0.04}{\varepsilon_{ps} - 0.007} & \text{for } \varepsilon_{ps} > 0.0086 \end{cases} \quad (R7.3.6b)$$

R7.3.7 In unbonded prestressed members, the stress in the prestressing steel seldom exceeds yield; limiting the permitted design stress to $0.95 f_{py}$ allows the use of a linear stress-strain relationship for the unbonded prestressing steel.

R7.3.8.1 A description of the calibration of Eq. (7.3.8.2) is provided in ACI PRC-440.2.

CODE

R7.3.8.2 A description of the calibration of Eq. (7.3.8.3) is provided in ACI PRC-440.2.

7.4—Nominal flexural strength

7.4.1 This section shall apply to reinforced concrete members, prestressed concrete members having bonded prestressing reinforcement, and prestressed concrete members having unbonded prestressing reinforcement strengthened in flexure by externally bonded or NSM FRP reinforcement applied at the tension face of the member.

7.4.2 The stress in nonprestressed steel reinforcement under service load $f_{s,s}$ shall not exceed $0.80 f_y$.

7.4.3 The compressive stress in concrete under service load $f_{c,s}$ shall not exceed $0.60 f_c'$.

7.4.4 The stress in prestressed steel reinforcement under service load $f_{ps,s}$ shall not exceed $0.94 f_{py}$ or $0.80 f_{pu}$.

7.4.5 The effective strains in the internal nonprestressed steel reinforcement, prestressed steel reinforcement, and FRP reinforcement shall be determined using strain compatibility in accordance with 7.3.4.

COMMENTARY

R7.4—Nominal flexural strength

R7.4.4 To avoid inelastic deformations of the strengthened member, the prestressing steel should be prevented from yielding under service load levels. The stress in the prestressing steel can be calculated based on the actual condition (cracked or uncracked section) of the strengthened reinforced concrete section. The strain in prestressing steel at service, $\varepsilon_{ps,s}$, can be calculated from Eq. (R7.4.4a).

$$\varepsilon_{ps,s} = \varepsilon_{pe} + \frac{P_e}{A_c E_c}\left(1 + \frac{e^2}{r^2}\right) + \varepsilon_{pnet,s} \quad (R7.4.4a)$$

The net tensile strain in the prestressing steel beyond decompression at service $\varepsilon_{pnet,s}$ depends on the effective section properties at service, and can be calculated using Eq. (R7.4.4b) or (R7.4.4c).

For uncracked section at service: $\varepsilon_{pnet,s} = \dfrac{M_s e}{E_c I_g}$ (R7.4.4b)

For cracked section at service: $\varepsilon_{pnet,s} = \dfrac{M_{snet} e}{E_c I_{cr}}$ (R7.4.4c)

R7.4.5 Figure R7.4.5 illustrates the internal strain and stress distribution for a singly reinforced nonpresstressed rectangular section under flexure at the ultimate limit state.

The calculation procedure used to arrive at the ultimate strength should satisfy strain compatibility and force equilibrium, and should consider the governing mode of failure. For any assumed depth to the neutral axis, c, the strain in the FRP reinforcement can be calculated from Eq. (R7.4.5a).

$$\varepsilon_{fe} = \varepsilon_{cu}\left(\frac{d_f - c}{c}\right) - \varepsilon_{bi} \leq \varepsilon_{fd} \quad (R7.4.5a)$$

Equation (R7.4.5a) considers the governing mode of failure for the assumed neutral axis depth. If the left term of

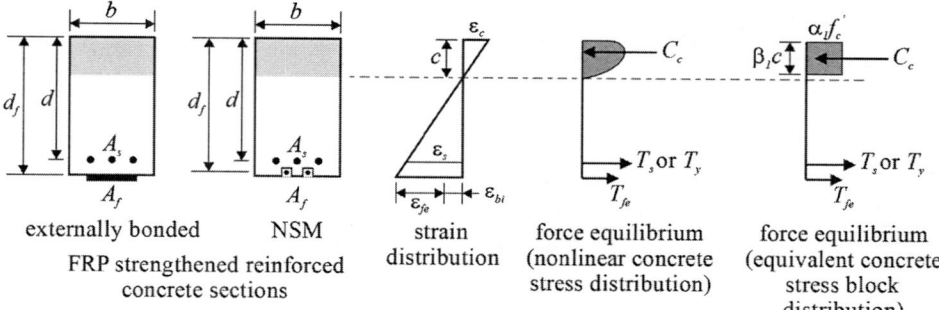

Fig. R7.4.5—Internal strain and stress distribution for a singly-reinforced rectangular section under flexure at ultimate limit state.

the inequality controls, concrete crushing controls flexural failure of the section. If the right term of the inequality controls, FRP failure (rupture or debonding) controls.

The initial strain ε_{bi} is the strain in the substrate at the time of application of the FRP. This will typically include the effects of member dead load and additional loads not relieved during the strengthening application. The calculation procedure for the initial strain ε_{bi} will depend on the state of the concrete section at the time of FRP installation and at service condition. Nonpresstressed sections will be cracked. Prestressed sections can be uncracked at installation/uncracked at service, uncracked at installation/cracked at service, or cracked at installation/cracked at service.

The initial strain on the bonded substrate, ε_{bi}, can be determined from an elastic analysis of the existing member, considering all loads that will be present during the installation of the FRP system. The elastic analysis of the existing member should be based on cracked or uncracked section properties, depending on existing conditions.

The strain in the nonprestressed steel reinforcement can be calculated from Eq. (R7.4.5b).

$$\varepsilon_s = (\varepsilon_{fe} + \varepsilon_{bi})\left(\frac{d-c}{d_f-c}\right) \quad \text{(R7.4.5b)}$$

Similarly, for a prestressed member, the strain in the FRP reinforcement can be calculated from Eq. (R7.4.5c)

$$\varepsilon_{fe} = (\varepsilon_{pu} - \varepsilon_{pi})\left(\frac{d_f-c}{d_p-c}\right) - \varepsilon_{bi} \leq \varepsilon_{fd} \quad \text{(R7.4.5c)}$$

Where the initial strain in prestressed steel reinforcement can be calculated from Eq. (R7.4.5d).

$$\varepsilon_{pi} = \frac{P_e}{A_p E_s} + \frac{P_e}{A_c E_c}\left(1 + \frac{e^2}{r^2}\right) \quad \text{(R7.4.5d)}$$

The strain in the prestressed steel reinforcement can be calculated from Eq. (R7.4.5e).

CODE

7.4.6 The effective stresses in nonprestressed internal steel reinforcement, prestressed steel reinforcement and FRP reinforcement shall be determined in accordance with 7.3.5, 7.3.6, and 7.3.8, respectively.

7.4.7 The effective stresses in unbonded prestressed steel reinforcement shall be determined in accordance with 7.3.7 in which the effective strain shall be calculated by Eq. (7.4.7).

$$\varepsilon_{ps} = \varepsilon_{pe} + \eta \varepsilon_c \left(\frac{d_p - c}{L_a} \right) \quad (7.4.7)$$

7.4.7.1 The factor η shall be determined in accordance with Table 7.4.7.1.

7.4.8 The stress in the FRP under service loads $f_{f,s}$ shall be limited to the values given in 6.5.1.

COMMENTARY

$$\varepsilon_{ps} = \varepsilon_{pe} + \frac{P_e}{A_c E_c} \left(1 + \frac{e^2}{r^2} \right) + \varepsilon_{pnet} \leq 0.035 \quad (R7.4.5e)$$

where ε_{pnet} is the net tensile strain in the prestressing steel beyond decompression at the nominal strength. The value of ε_{pnet} will depend on the mode of failure and can be calculated using Eq. (R7.4.5f) or (R7.4.5g).

For concrete crushing failure:

$$\varepsilon_{pnet} = 0.003 \left(\frac{d_p - c}{c} \right) \quad (R7.4.5f)$$

For FRP rupture or debonding failure modes:

$$\varepsilon_{pnet} = (\varepsilon_{fe} + \varepsilon_{bi}) \left(\frac{d_p - c}{d_f - c} \right) \quad (R7.4.5g)$$

R7.4.7 In sections having unbonded prestressed reinforcement, the unbonded steel slips relative to the surrounding concrete resulting in the calculation of steel strain or stress becoming a function of overall member deformation rather than only section curvature. The same equilibrium approach applied to a flexural section with bonded prestressing may be used to determine the nominal strength of the FRP-strengthened member, provided an appropriate method is used to calculate the strains or stresses in the unbonded tendons at the ultimate flexural strength (El Meski and Harajli 2014; ACI PRC-440.2). In Eq. (7.4.7), ε_{pe} is the effective strain in the unbonded prestressing steel after losses, and La is the total length of tendon between anchorages.

R7.4.7.1 The parameter η combines the effects of member continuity and applied load pattern for producing maximum factored moment at the critical section under consideration (Harajli 2012) ACI PRC-440.2).

R7.4.8 For nonprestressed members, the stress in the FRP reinforcement can be calculated from Eq. (R7.4.8a).

Table 7.4.7.1—Values of factor η

Positive moment			Negative moment			
Simply supported spans	Exterior spans	Interior spans	Two span members, first interior support	Three or more spans, first interior support	All other interior supports	Cantilever spans
14.0	19.0	24.5	38.5	43.5	49.0	5.3

CODE

COMMENTARY

$$f_{f,s} = f_{s,s}\left(\frac{E_f}{E_s}\right)\frac{d_f - kd}{d - kd} - \varepsilon_{bi}E_f \quad (R7.4.8a)$$

The stress in the FRP reinforcement under an applied moment within the elastic response range of a prestressed member can be calculated from Eq. (R7.4.8b).

$$f_{f,s} = \left(\frac{E_f}{E_c}\right)\frac{M_s y_b}{I_g} - \varepsilon_{bi}E_f \quad (R7.4.8b)$$

7.4.9 The nominal flexural strength of the section with FRP reinforcement determined from equilibrium at a section shall include an additional strength reduction factor $\psi_f = 0.85$, applied to the FRP reinforcement contribution to flexural strength.

R7.4.9 The nominal flexural strength of a singly reinforced rectangular section with FRP reinforcement can be calculated from Eq. (R7.4.9).

$$M_n = A_s f_s\left(d_s - \frac{\beta_1 c}{2}\right) + A_{ps}f_{ps}\left(d_{ps} - \frac{\beta_1 c}{2}\right) + \psi_f A_f f_{fe}\left(d_f - \frac{\beta_1 c}{2}\right)$$
(R7.4.9)

The reduction factor $\psi_f = 0.85$ is based on a reliability analysis based on the experimentally calibrated statistical properties of the flexural strength (ACI PRC-440.2).

7.5—Moment redistribution for continuous reinforced concrete beams

7.5.1 Moment redistribution shall be permitted when the strain in the tension steel reinforcement, ε_t, exceeds 0.0075 at the section where moment is reduced.

7.5.2 Moment redistribution shall not be permitted where the moments have been calculated using the approximate method described in ACI CODE-318-19 Section 6.5.

7.5.3 The reduction of maximum negative or maximum positive moment for any assumed loading arrangement shall not exceed the lesser of $1000\varepsilon_t$ percent and 20%.

7.5.4 Static equilibrium shall be maintained after redistribution of moments for each loading arrangement.

7.6—Anchorage and development of externally bonded FRP

7.6.1 Concrete substrate surface preparation shall be in accordance with 5.2.

7.6.2 Anchorage and detailing at the termination of externally bonded FRP for flexural strengthening shall conform to 7.6.2.1, 7.6.2.2, or 7.6.2.3.

7.6.2.1 For simply supported beams, all FRP laminates shall be terminated a distance greater than ℓ_{df}, determined in accordance with 7.6.3, past the point along the span at which

R7.5—Moment redistribution for continuous reinforced concrete beams

R7.5.3 El-Refaie et al. (2003) demonstrated that continuous reinforced concrete beams strengthened with carbon FRP sheets can redistribute moment in the order of 6 to 31%.

R7.6—Anchorage and development of externally bonded FRP

R7.6.2 The requirements of this section are intended to mitigate the "end peel debonding" mode of failure (ACI PRC-440.2) and are illustrated in Fig. R7.6.2.

R7.6.2.1 Additional robustness can be provided by extending flexural FRP as close as possible to the point of zero moment in a simply supported span (ACI PRC-440.2).

CODE

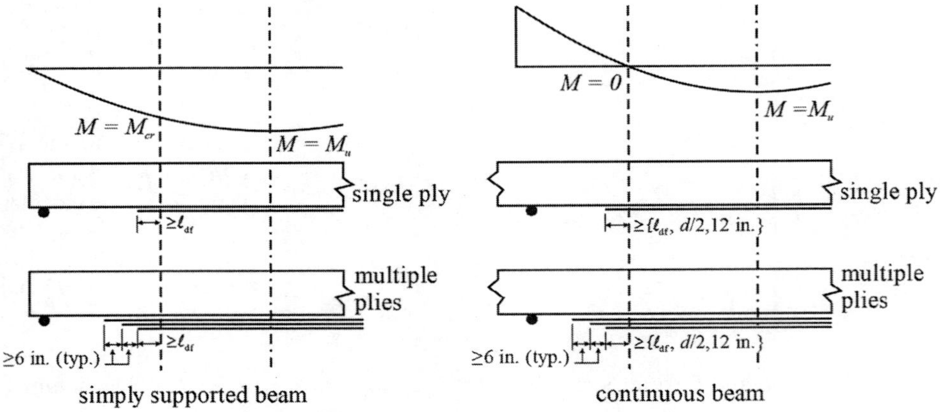

Fig. R7.6.2—Termination requirements for externally bonded FRP flexural strengthening.

the resisted moment falls below the cracking moment M_{cr} of the beam.

7.6.2.2 For continuous beams, all FRP laminates shall be terminated a distance greater than ℓ_{df} and $d/2$ and 12 in. beyond the inflection point.

7.6.2.3 When the requirements of 7.6.2.1 or 7.6.2.2 cannot be met or the factored shear force in the member exceeds two-thirds of the concrete shear strength ($V_u > 0.67V_c$) at the termination of flexural FRP, the flexural FRP shall be anchored with transverse U-wrap anchorage having an area given by Eq. (7.6.2.3).

$$A_{fanchor} = \frac{(A_f f_{fe})_{flexural\ FRP}}{(E_f \kappa_v \varepsilon_{fu})_{anchor}} \qquad (7.6.2.3)$$

7.6.2.3.1 The bond-reduction coefficient κ_v shall be determined from Eq. (8.6.5.3).

7.6.2.3.2 The U-wrap anchorage shall extend vertically up the sides of the web a distance equal to the lesser of the depth of the web and 12 in.

7.6.2.3.3 The U-wrap anchorage shall extend a distance greater than ℓ_{df} and $d/2$ and 12 in. from the termination of the flexural FRP.

7.6.3 The development length ℓ_{df} of externally bonded FRP for flexural strengthening shall be determined from Eq. (7.6.3).

$$\ell_{df} = 0.057\sqrt{NE_f t_f / \sqrt{f'_c}} \qquad (7.6.3)$$

7.6.4 For multiple-ply FRP systems, the terminations of the plies shall conform to the requirements of 7.6.4.1 and 7.6.4.2.

COMMENTARY

R7.6.4 Providing a stepwise termination of successive FRP plies reduces the stress concentrations responsible for end-peel debonding. The requirements of 7.6.4 are illustrated in Fig. R7.6.2.

CODE

7.6.4.1 The outermost ply shall be terminated in accordance with 7.6.2.

7.6.4.2 Successive plies beneath the outermost ply shall be longer and terminated no less than 6 in. from the end of the previous ply.

7.6.5 Lap splices in the longitudinal direction of wet layup FRP shall have a splice length greater than ℓ_{df} determined in accordance with 7.6.3 and 6 in.

7.6.5.1 Splicing of precured laminates shall not be permitted.

7.7—Development of near-surface-mounted FRP flexural strengthening

7.7.1 Concrete substrate surface preparation and groove preparation shall be in accordance with 5.4.

7.7.2 Grooves into which near-surface-mounted (NSM) FRP is placed shall conform to 7.7.2.1 or 7.7.2.2.

7.7.2.1 For circular NSM FRP having diameter d_b, both the groove depth and width shall equal or exceed $1.5d_b$.

7.7.2.2 For rectangular NSM FRP having dimensions $a_b \times b_b$ in which a_b is the smaller dimension, groove depth shall equal or exceed $1.5b_b$ and groove width shall equal or exceed $3a_b$.

7.7.3 Clear spacing between adjacent NSM grooves shall exceed twice the groove depth.

7.7.4 Clear spacing between NSM grooves and the edge of a member shall exceed four times the groove depth.

COMMENTARY

R7.6.5 For unidirectional wet layup FRP laminates, lap splices are required only in the direction of the fibers. Lap splices are not required in the direction transverse to the fibers. Lap splices should be located outside of regions of high expected FRP stress and should be staggered to the extent possible.

R7.7—Development of near-surface-mounted FRP flexural strengthening

R7.7.2 Minimum groove dimensions are shown in Fig. R7.7.2. Care should be taken that groove depth does not impinge upon existing internal reinforcing steel. Grooves should be cleaned in accordance with manufacturers requirements.

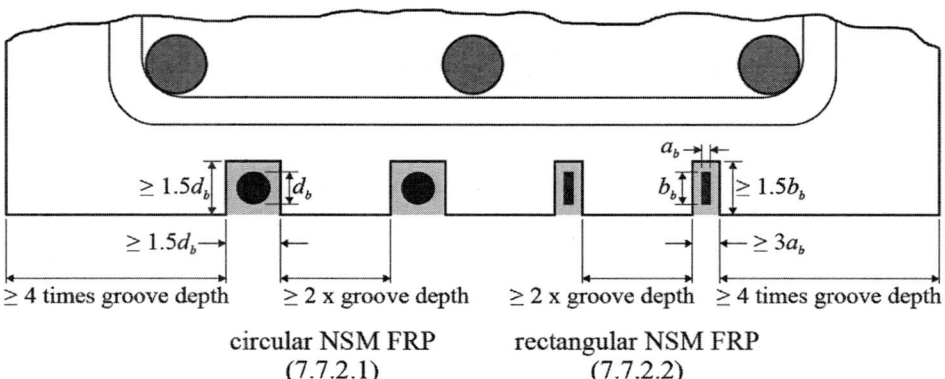

Fig. R7.7.2—Minimum dimensions of NSM grooves.

CODE

7.7.5 Termination of NSM FRP shall conform to 7.6.2.1 or 7.6.2.2, where ℓ_{df} is determined in accordance with 7.7.6.

7.7.6 The development length ℓ_{df} of NSM FRP for flexural strengthening shall be determined from Eq. (7.7.6a and b). For circular NSM FRP having diameter d_b:

$$\ell_{df} = \frac{f_{fd} d_b}{4\tau_b} \quad (7.7.6a)$$

For rectangular NSM FRP having dimensions $a_a \times b_b$:

$$\ell_{df} = \frac{f_{fd} a_b b_b}{4\tau_b (a_b + b_b)} \quad (7.7.6b)$$

7.7.6.1 The bond strength τ_b shall be taken as 1000 psi.

COMMENTARY

CODE

CHAPTER 8—DESIGN AND DETAILING FOR SHEAR STRENGTHENING

8.1—General

8.1.1 This chapter shall apply to the design and detailing of shear strengthening of reinforced concrete and prestressed concrete members using externally bonded FRP systems when permitted by the provisions of this Code.

8.2—Sectional requirements

8.2.1 Columns strengthened for shear shall be rectangular or circular in section except as permitted by 8.2.1.1.

8.2.1.1 Shear strengthening of polygonal columns having more than four sides and elliptical columns is permitted using the provisions for a circular section and an effective diameter equal to the diameter of the circle completely inscribing the actual cross section.

8.3—FRP System wrapping schemes

8.3.1 FRP strengthening for shear shall be designated as being bond critical.

8.3.2 The FRP system wrapping scheme for beams shall be designated as being completely wrapped, U-wrapped, or two-sided in accordance with Fig. 8.3.2.

COMMENTARY

CHAPTER R8—DESIGN AND DETAILING FOR SHEAR STRENGTHENING

R8.1—General

R8.1.1 Additional strength provided by FRP systems is a function of many factors, including beam geometry, wrapping scheme, and existing concrete strength.

8.3—FRP System wrapping schemes

R8.3.2 Complete wrapping of the beam section is the most efficient means of shear strengthening followed by U-wraps. FRP bonded to only two opposing sides of the section is the least efficient scheme.

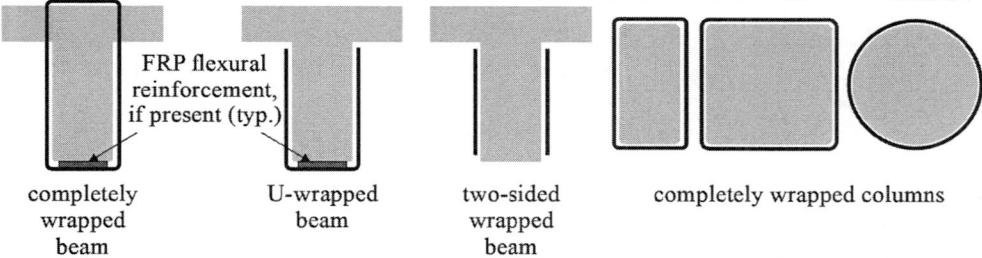

Fig. 8.3.2—Permitted FRP wrapping schemes for shear strengthening beams and columns.

8.3.3 For columns, the FRP system shall be completely wrapped around the section and shall be oriented such that the fibers are oriented perpendicular to the longitudinal axis of the member.

8.3.4 At sections of a member where FRP flexural reinforcement is also present, FRP shear reinforcement shall be placed after the flexural FRP.

8.3.5 For completely wrapped sections, each discrete sheet of FRP shall have an overlapped closure splice greater than ℓ_{df} determined in accordance with 7.6.3 calculated with $N = 1$ and 6 in. long.

8.3.6 For rectangular sections, the overlapped closure splice shall be located such that no part of splice rounds an exterior corner.

R8.3.4 Shear reinforcement placed on top of flexural reinforcement, as shown in Fig. 8.3.2, is known to provide some anchorage to the flexural reinforcement (ACI PRC-440.2).

R8.3.5 Multiple plies of FRP can be provided using a single continuous sheet in accordance with 8.3.5. Manufacturer recommendations may require a longer overlap splice.

CODE

8.3.7 It is permitted to install the FRP shear reinforcement continuously or in discrete strips along the length of the member over which it is required.

8.3.7.1 For discrete FRP strips, the clear spacing between adjacent strips shall not exceed the lesser of $d/4$ and 12 in.

8.4—Design strength

8.4.1 The design strength at all sections shall satisfy Eq. (8.4.1).

$$\phi V_n \geq V_u \quad (8.4.1)$$

8.4.2 The strength reduction factor ϕ shall be determined in accordance with ACI CODE-562.

8.5—Nominal shear strength

8.5.1 The nominal shear strength of an FRP-strengthened member shall be determined in accordance with Eq. (8.5.1).

$$V_n = V_c + V_s + \psi_f V_f \quad (8.5.1)$$

8.5.2 The concrete and internal steel contributions to nominal shear strength, V_c and V_s, respectively, shall be computed in accordance with the design basis code.

8.5.3 The strength reduction factor ψ_f shall be taken as:
(a) 0.95 for completely wrapped members
(b) 0.85 for U-wraps and two-sided FRP applications

8.5.4 Total shear strength provided by the sum of the internal steel and FRP shear reinforcement components shall not exceed the limit given by Eq. (8.5.4).

For rectangular sections: $V_s + V_f \leq 8\sqrt{f'_c} b_w d$ (in.-lb) (8.5.4)

8.5.4.1 For circular sections, $b_w = D_c$ and $d = 0.8 D_c$ shall be substituted in Eq. (8.5.4).

8.6—FRP Contribution to shear strength

8.6.1 The FRP contribution to shear strength shall be calculated by Eq. (8.6.1).

$$V_f = \frac{A_{fv} f_{fe} d_{fv} (\sin\alpha + \cos\alpha)}{s_f} \quad (8.6.1)$$

8.6.2 The effective depth of FRP strengthening d_{fv} shall be in accordance with Fig. 8.6.2 for rectangular sections and shall be 0.8 times the diameter of the section for circular sections.

COMMENTARY

R8.3.7 When FRP systems are installed continuously along a member, the potential adverse effects of entrapping moisture in the members may need to be considered as required by 6.7.1.

R8.5—Nominal shear strength

R8.5.3 The additional strength reduction factor ψ_f for U-wraps and two-sided FRP applications is based on reliability analysis described in ACI PRC-440.2. U-wraps and two-sided applications are more susceptible to debonding at their terminations whereas complete wrapping provides greater reliability warranting a higher value of ψ_f (ACI PRC-440.2).

R8.5.4 This limit is intended to minimize the likelihood of diagonal compression failure in the concrete and to limit the extent of cracking.

R8.6—FRP Contribution to shear strength

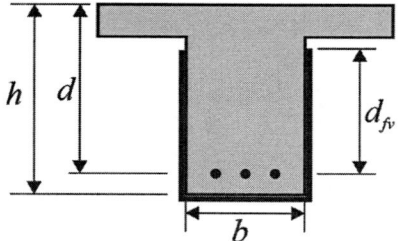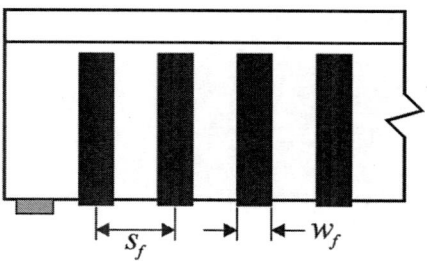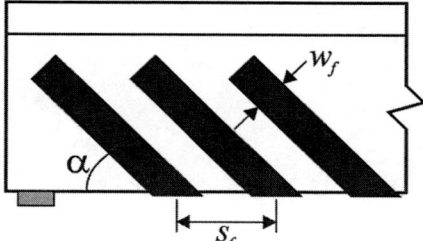

Fig. 8.6.2—Dimension definitions of FRP shear strengthening.

8.6.3 The angle of inclination of FRP strips relative to the longitudinal axis of the member α shall be between 45 degrees and 90 degrees, inclusive.

8.6.4 The area of FRP shear reinforcement shall be determined in accordance with Eq. (8.6.4a) or (8.6.4b).

For rectangular sections: $A_{fv} = 2Nt_f w_f$ (8.6.4a)

For circular sections: $A_{fv} = (\pi/2)Nt_f w_f$ (8.6.4b)

8.6.5 The effective strain in the FRP, ε_{fe}, shall be determined in accordance with 8.6.5.1 or 8.6.5.2.

R8.6.5 The effective strain is the maximum strain that can be achieved in the FRP at the nominal strength of the member and is a function of the existing concrete compressive strength and failure mode of the FRP.

8.6.5.1 For completely wrapped members and for U-wraps anchored in accordance with 8.7, the effective strain ε_{fe} shall be determined from Eq. (8.6.5.1).

$$\varepsilon_{fe} = 0.004 \leq 0.75\varepsilon_{fu} \quad (8.6.5.1)$$

R8.6.5.1 The strain limit of 0.004 for completely wrapped members is intended to ensure the continued concrete contribution, V_c, to the nominal shear strength through the maintenance of aggregate interlock. The value 0.004 is based on testing and experience (ACI PRC-440.2).

8.6.5.2 For U-wraps and two-sided FRP applications, the effective strain ε_{fe} shall be determined from Eq. (8.6.5.2).

$$\varepsilon_{fe} = \kappa_v \varepsilon_{fu} \leq 0.004 \quad (8.6.5.2)$$

R8.6.5.2 For U-wraps and two-sided FRP applications, the FRP has been observed to delaminate from the concrete prior to the degradation of the concrete contribution, V_c, to the nominal shear strength. The bond-reduction coefficient κ_v is a function of concrete strength, wrapping scheme, and stiffness of the FRP system. The design procedure adopted herein was developed by a combination of analytical and empirical results (ACI PRC-440.2).

8.6.5.3 The bond-reduction coefficient κ_v shall be determined from Eq. (8.6.5.3).

$$\kappa_v = \frac{k_1 k_2 L_e}{468\varepsilon_{fu}} \leq 0.75 \quad (8.6.5.3)$$

8.6.5.3.1 The active bond length L_e shall be calculated from Eq. (8.6.5.3.1).

$$L_e = \frac{2500}{(Nt_f E_f)^{0.58}} \quad (8.6.5.3.1)$$

CODE

8.6.5.3.2 The modification factor k_1 shall be calculated from Eq. (8.6.5.3.2).

$$k_1 = \left(\frac{f_c'}{4000}\right)^{2/3} \quad (8.6.5.3.2)$$

8.6.5.3.3 The modification factor k_2 shall be calculated from Eq. (8.6.5.3.3a) or Eq. (8.6.5.3.3b), as applicable.

For U-wraps: $k_2 = (d_{fv} - L_e)/d_{fv}$ \quad (8.6.5.3.3a)

For two-sided FRP: $k_2 = (d_{fv} - 2L_e)/d_{fv}$ \quad (8.6.5.3.3b)

8.7—Anchorage for U-wraps

8.7.1 Fiber anchors used to anchor U-wrap shear reinforcement such that effective strain in accordance with 8.6.5.1 can be achieved shall conform to all requirements of 8.7.

8.7.2 U-wraps anchored by fiber anchors shall have laminate stiffness, $Nt_f E_f$, no greater than 1150 kip/in.

8.7.3 For continuous U-wraps, fiber anchors shall be spaced (s_{anc}) along the longitudinal axis of the member at a distance no greater than 10 in.

8.7.4 For discrete U-wraps, each U-wrap shall be anchored and the anchor spacing on a single discrete U-wrap, s_{anc}, shall not exceed 10 in.

8.7.5 Tensile strength and elastic modulus of the FRP material from which fiber anchor is fabricated shall be equal to or greater than those of the U-wraps being anchored.

8.7.6 The gross laminate area of the fiber anchor securing each leg of the U-wrap shall be determined by Eq. (8.7.6).

$$A_{anc} \geq R_A(Nt_f s_{anc}) \quad (8.7.6)$$

8.7.6.1 The factor R_A shall be determined in accordance with Table 8.7.6.1

8.7.6.2 For discrete U-wraps having a single anchor, s_{anc} in Eq. (8.7.6) shall be taken as the width of the discrete strip, w_f.

COMMENTARY

R8.7—Anchorage for U-wraps

R8.7.1 U-wraps anchored with fiber anchors have been shown to increase the shear strength of full-scale concrete bridge girders by 20 to 60% and to demonstrate favorable fatigue and long-term creep performance (Jirsa et al. 2017). The limitations of the anchorage approach described in 8.7 are based on available testing. Additional information on fiber anchors for U-wraps can be found in ACI PRC-440.2.

R8.7.4 Discrete U-wraps having a width w_f up to 10 in. can be anchored with a single fiber anchor. Wider discrete U-wraps require multiple anchors having a spacing not exceeding 10 in.

R8.7.6 The required area of fibers in a fiber anchor is a function of force to be developed in the U-wrap, the splay angle α_{anc}, and the embedment angle β_{anc} (del Rey Castillo et al. 2019; ACI PRC-440.2). A_{anc} is the minimum area of a saturated fiber anchor with an amount of fiber equal to or greater than the fiber in a width of the laminate represented by $R_A N s_{anc}$. The effectiveness of a fiber anchor diminishes with an increase in anchor cross-sectional area and increase in width of U-wrap anchored (Pudleiner et al. 2019; ACI PRC-440.2).

CODE

Table 8.7.6.1—Fiber anchor design parameters and detailing requirements, in.-lb

s_{anc}, in.	U-wrap NE_ft_f, kip/in.	R_A 90 degrees $\leq \beta_{anc} \leq 125$ degrees	R_A 125 degrees $< \beta_{anc} \leq 180$ degrees	r_{anc}, in.	h_{anc}, in. 90 degrees $\leq \beta_{anc} \leq 110$ degrees	h_{anc}, in. 110 degrees $< \beta_{anc} \leq 125$ degrees	h_{anc}, in. 125 degrees $< \beta_{anc} \leq 180$ degrees
≤ 4	$NE_ft_f \leq 288$	1.25	1.00	s_{anc}			6.0
	$288 < NE_ft_f \leq 575$	1.25	1.00	s_{anc}			6.0
	$575 < NE_ft_f \leq 863$	1.25	1.00	8.0			6.0
	$863 < NE_ft_f \leq 1150$	1.25	1.00	10.0			8.0
$4 < s_{anc} \leq 6$	$NE_ft_f \leq 288$	1.25	1.00	s_{anc}			6.0
	$288 < NE_ft_f \leq 575$	1.25	1.00	s_{anc}			6.0
	$575 < NE_ft_f \leq 863$	1.50	1.25	8.0	Larger of 4.0 and $7(A_{anc})^{0.5}$	Larger of 6.0 and $7(A_{anc})^{0.5}$	8.0
	$863 < NE_ft_f \leq 1150$	1.50	1.25	10.0			10.0
$6 < s_{anc} \leq 8$	$NE_ft_f \leq 288$	1.25	1.00	s_{anc}			6.0
	$288 < NE_ft_f \leq 575$	1.50	1.25	s_{anc}			8.0
	$575 < NE_ft_f \leq 863$	1.50	1.25	s_{anc}			10.0
	$863 < NE_ft_f \leq 1150$	1.75	1.50	10.0			12.0
$8 < s_{anc} \leq 10$	$NE_ft_f \leq 288$	1.25	1.00	s_{anc}			6.0
	$288 < NE_ft_f \leq 575$	1.50	1.25	s_{anc}			8.0
	$575 < NE_ft_f \leq 863$	1.75	1.50	s_{anc}			10.0
	$863 < NE_ft_f \leq 1150$	2.00	2.00	s_{anc}			12.0

8.7.7 Fiber anchors shall be detailed in accordance with 8.7.7.2 or 8.7.7.3.

COMMENTARY

R8.7.7 The detailing requirements of 8.7.7 and 8.7.8 are essential to achieving the desired anchor performance and required effective strain in the anchored U-wraps. Figure R8.7.7 illustrates the requirements of 8.7.7 and 8.7.8 including the determination of d_{fv} described in 8.7.7.1.

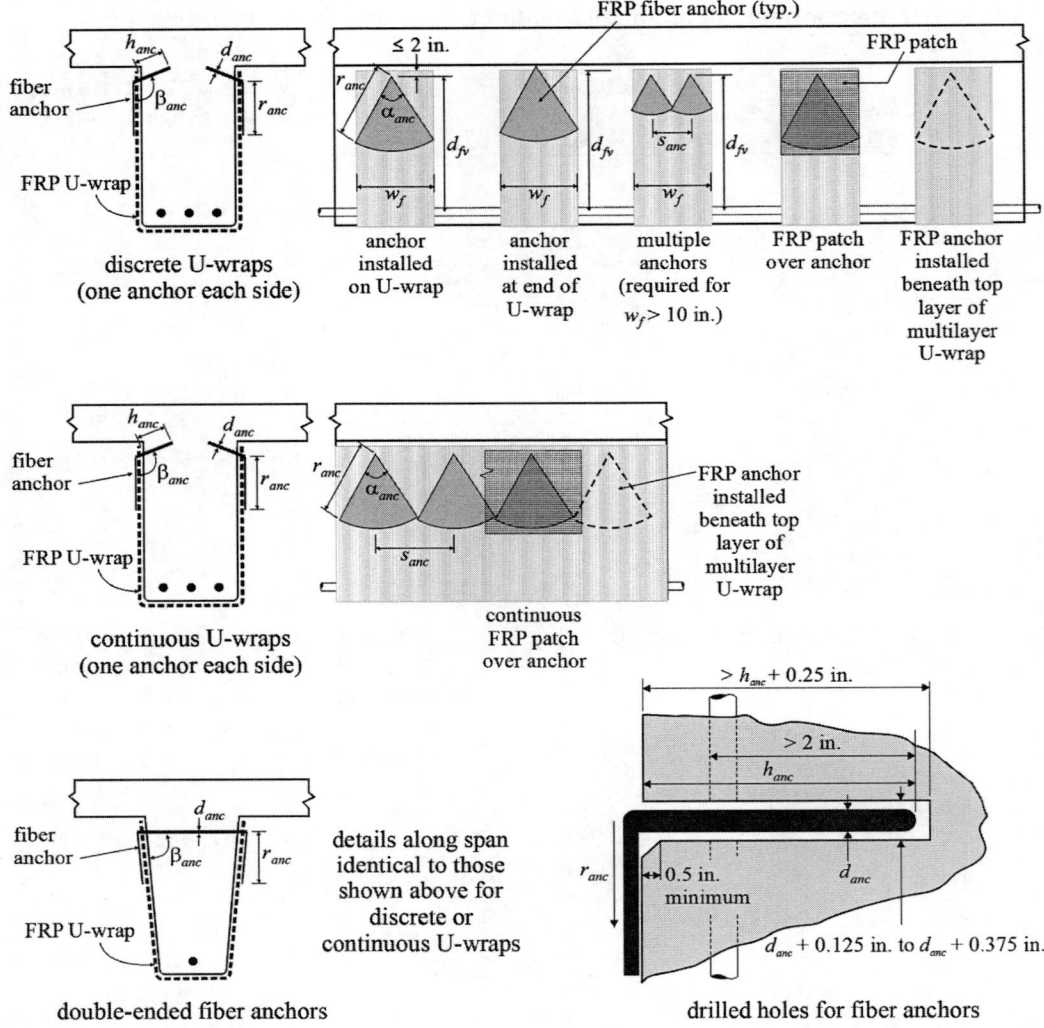

Fig. R8.7.7—Fiber anchors for anchoring U-wraps.

8.7.7.1 For anchored U-wraps, d_{fv} shall be measured from the location of the anchor.

8.7.7.2 Individual fiber anchors having a single splay shall be installed on both legs of the U-wrap being anchored at a location either immediately above the end of the U-wrap or on the U-wrap no further than 2 in. from the end of the U-wrap. Individual fiber anchors shall be detailed in accordance with (a) through (f).

(a) Fiber shall be distributed uniformly across the splay and need not extend beyond the edge of the U-wrap.
(b) Splay angle α_{anc} shall not exceed 60 degrees.
(c) Splay length r_{anc} shall be determined in accordance with Table 8.7.6.1.
(d) If the anchor is not finished in accordance with 8.7.8, the anchor splay length r_{anc} provided shall be 1.5 times that determined in accordance with Table 8.7.6.1.
(e) Embedment angle β_{anc} shall not be less than 90 degrees or greater than 180 degrees.

CODE

(f) Embedment depth h_{anc} shall be determined in accordance with Table 8.7.6.1 and shall extend at least 2 in. beyond concrete cover.

8.7.7.3 Fiber anchors extending through the width of the cross section having a splay at both ends shall be installed such that both legs of the U-wrap are anchored at locations either immediately above the end of the U-wrap or on the U-wrap no further than 2 in. from the end of the U-wrap. Double-ended fiber anchors shall be detailed in accordance with (a) through (f).
 (a) Fiber shall be distributed uniformly across the splay and need not extend beyond the edge of the U-wrap.
 (b) Splay angle α_{anc} shall not exceed 60 degrees.
 (c) Splay length r_{anc} shall be determined in accordance with Table 8.7.6.1 if the anchor is finished in accordance with 8.7.8.
 (d) If the anchor is not finished in accordance with 8.7.8, the anchor splay length, r_{anc}, provided shall be 1.5 times that determined in accordance with Table 8.7.6.1.
 (e) The embedment angle β_{anc} for double ended fiber anchors shall be taken as 90 degrees.
 (f) The anchor shall be placed through a single hole drilled through the member being strengthened.

8.7.8 The fiber anchor shall be finished in accordance with 8.7.8.1 or 8.7.8.2.

8.7.8.1 For U-wraps composed of multiple FRP plies, the fiber anchor shall be installed under the topmost U-wrap layer.

8.7.8.2 A single layer patch of the same FRP system as the U-wrap shall be installed over the anchor splay with the FRP oriented perpendicular to the anchored U-wrap.

8.7.9 The drilled holes into which the fiber anchors are installed shall be detailed in accordance with (a) through (d):
 (a) Hole drilling and cleaning shall be in accordance with the fiber anchor manufacturer's installation requirements.
 (b) The drilled hole diameter shall be between 0.125 in. and 0.375 in. larger than the anchor diameter, d_{anc}, corresponding to A_{anc}.
 (c) The drilled hole depth shall be no less than 0.25 in. deeper than h_{anc}.
 (d) The edge of the drilled hole shall be chamfered or rounded to a distance no less than 0.5 in. over the extent of the splay angle α_{anc}.

8.8—Details for FRP shear strengthening

8.8.1 Concrete substrate surface preparation shall be in accordance with 5.2.

8.8.2 FRP shall not turn inside corners without the provision of anchors, designed in accordance with 8.7, provided on both sides of the inside corner.

COMMENTARY

R8.7.7.3 Double ended through-width fiber anchors may be used in place of individual fiber anchors on each side of the U-wrap. Such anchors may be used for members having relatively thin webs such as cast-in-place pan joints and precast concrete tees.

R8.7.9 The rounding of the hole edge (d) is intended to reduce stress concentrations at the anchor bend.

R8.8—Details for FRP shear strengthening

R8.8.2 Examples of inside corners are the intersections of beams and joists and the intersection of sides of beams and the underside of slabs.

CODE

CHAPTER 9—DESIGN AND DETAILING FOR AXIAL FORCE AND COMBINED AXIAL FORCE AND MOMENT STRENGTHENING

9.1—General

9.1.1 This chapter shall apply to the design and detailing of axial and combined axial and flexural strengthening of reinforced concrete members when permitted by the provisions of this Code.

9.1.2 The provisions of this chapter shall not apply to columns with unconfined concrete strength, f_c', greater than 10,000 psi.

9.1.3 The provisions of this chapter shall apply only to FRP systems in full contact with flat or convex surfaces.

9.1.4 The provisions of this chapter shall not apply to rectangular columns with any side dimension greater than 36 in.

9.1.5 The provisions of this chapter shall not apply to rectangular columns having a ratio of side dimensions less than 0.5 or greater than 2.0.

9.1.6 The provisions of this chapter shall apply to circular columns of any diameter.

9.2—Axial compression

9.2.1 The confinement of reinforced concrete columns shall be with completely wrapped FRP jackets.

9.2.1.1 FRP jackets are permitted to be designated as contact critical.

COMMENTARY

CHAPTER R9—DESIGN AND DETAILING FOR AXIAL FORCE AND COMBINED AXIAL FORCE AND MOMENT STRENGTHENING

R9.1—General

R9.1.1 Confinement of reinforced concrete compression members by means of FRP jackets can be used to enhance their strength and ductility, typically for extreme load conditions. An increase in axial-load-bearing capacity is calculated in terms of improved peak load resistance. Ductility enhancement requires more complex calculation to determine the ability of a member to sustain rotation and drift without a substantial loss in strength. This chapter applies only to reinforced concrete members confined with FRP systems.

R9.1.2 Strength enhancement for compression members with f_c' greater than 10,000 psi has not been experimentally verified. Effects of FRP confinement are known to become less efficient as unconfined concrete strength increases (ACI PRC-440.2).

R9.1.3 FRP confinement cannot be engaged when applied to a concave surface or over a large void or gap.

R9.1.4 The effectiveness of FRP confinement on rectangular columns having sides longer than 36 in. has not been experimentally verified. Confinement of long flat sides produces little enhancement of concrete behavior (Lam and Teng 2003b; ACI PRC-440.2).

R9.1.5 Confinement of rectangular columns with a large cross section aspect ratio produces only marginal increase in axial capacity (Lam and Teng 2003b; ACI PRC-440.2).

R9.2—Axial compression

FRP systems can be used to increase the axial compressive strength of a concrete member by providing confinement to the concrete in a manner similar to conventional spiral or tie reinforcing steel. Any contribution of longitudinally aligned fibers to the axial compression strength of a concrete member is beyond the scope of this Code. Strengthening for axial force alone is rare; strengthening for combined forces is addressed in 9.3.

R9.2.1.1 FRP jackets provide passive confinement to the compression member concrete, remaining unstressed until dilation and cracking of the wrapped compression member occur. For this reason, intimate contact between the FRP jacket and the concrete member is critical, although bond is not necessary and surface preparation may be in accordance with 5.3. If confinement reinforcement is also used for shear strengthening, the application is bond critical in

CODE

9.2.1.2 FRP jackets shall be completely wrapped around the section and shall be oriented such that the fibers are oriented perpendicular to the longitudinal axis of the member.

9.2.1.3 FRP jackets designed only to enhance strength shall be continuous over the length of the member being confined.

9.2.1.4 FRP jackets shall have an overlapped closure splice greater than 6 in. long located entirely on a flat side of a rectangular section.

9.2.2 The design strength at all sections shall satisfy Eq. (9.2.2).

$$\phi P_n \geq P_u \qquad (9.2.2)$$

9.2.2.1 The strength reduction factor ϕ shall be determined in accordance with ACI CODE-562.

9.2.3 The nominal axial compressive strength of a nonslender, nonprestressed, normalweight concrete member confined with an FRP jacket shall be determined in accordance with Eq. (9.2.3a) or (9.2.3b), As applicable.

For members with internal steel spiral reinforcement:

$$P_n = 0.85[0.85 f_{cc}'(A_g - A_{st}) + f_y A_{st}] \qquad (9.2.3a)$$

For members with internal steel-tie reinforcement:

$$P_n = 0.85[0.80 f_{cc}'(A_g - A_{st}) + f_y A_{st}] \qquad (9.2.3b)$$

9.2.4 The confined concrete compressive strength shall be determined in accordance with Eq. (9.2.4).

$$f_{cc}' = f_c' + \psi_f 3.3 \kappa_a f_\ell \qquad (9.2.4)$$

9.2.4.1 The maximum confining pressure shall be determined in accordance with Eq. (9.2.4.1a) or Eq. (9.2.4.1b), as applicable.

For a circular cross section having diameter D_c:

$$f_\ell = \frac{2 E_f N t_f \varepsilon_{fe}}{D_c} \qquad (9.2.4.1a)$$

For a rectangular cross section having dimensions $b_c \times h_c$:

$$f_\ell = \frac{2 E_f N t_f \varepsilon_{fe}}{\sqrt{b_c^2 + h_c^2}} \qquad (9.2.4.1b)$$

COMMENTARY

accordance with 8.3.1 and surface preparation according to 5.2 is required.

R9.2.1.3 FRP jackets intended to enhance ductility of a column may, in some instances, only be required over the length of the plastic hinges of the member (ACI PRC-440.2).

R9.2.1.4 A closure splice may be located anywhere around the circumference of a circular section.

CODE

9.2.4.1.1 The maximum effective strain ε_{fe} shall be $0.55\varepsilon_{fu}$.

9.2.4.1.2 The reduction factor ψ_f shall be taken as 0.95.

9.2.4.1.3 The ratio of f_ℓ to the unconfined concrete strength, f_c' shall be equal to or greater than 0.08.

COMMENTARY

R9.2.4.1.1 The FRP strain efficiency factor 0.55 accounts for the potential for premature FRP failure due to stress concentration caused by cracking of the confined concrete as it dilates (ACI PRC-440.2). Consideration should also be given to effects of sustained stress in the FRP as addressed in 6.5.

R9.2.4.1.3 Confinement corresponding to $f_\ell/f_c' = 0.08$ is the minimum level of confinement required to assure a nondescending post-peak branch in the stress-strain behavior of the confined concrete (Lam and Teng 2003a,b; ACI PRC-440.2). This is shown schematically in Fig. R9.2.4.1.3.

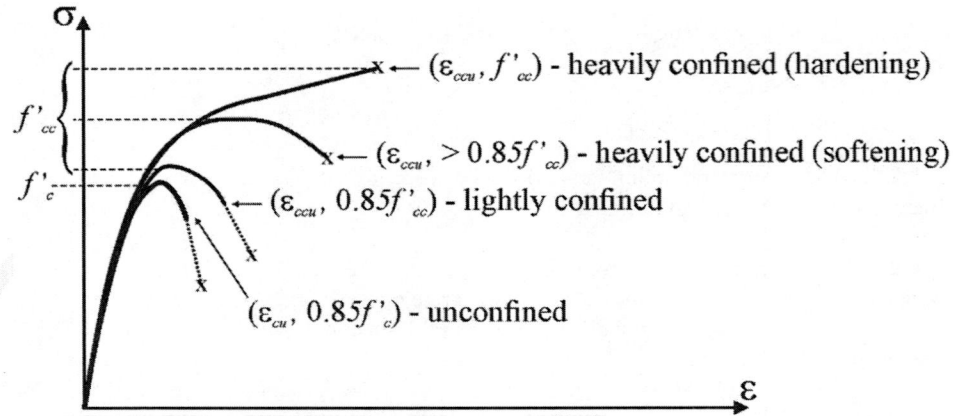

Fig. R9.2.4.1.3—Schematic stress-strain behavior of unconfined and confined reinforced concrete columns (after Rocca et al. [2006]).

9.2.4.2 For circular cross sections, the shape factor κ_a in Eq. (9.2.4) shall be taken as 1.0.

9.2.4.3 For rectangular cross sections having dimension $b_c \times h_c$, in which $h_c \leq b_c$, the shape factor κ_a in Eq. (9.2.4) shall be determined in accordance with Eq. (9.2.4.3).

$$\kappa_a = \frac{A_e}{A_g}\left(\frac{b_c}{h_c}\right)^2 \quad (9.2.4.3)$$

9.2.4.3.1 The ratio of effective confined area of concrete to gross concrete area, A_e/A_g shall be determined in accordance with Eq. (9.2.4.3.1).

$$\frac{A_e}{A_g} = \frac{1 - \dfrac{\left[\left(\dfrac{b_c}{h_c}\right)(h_c - 2r_c)^2 + \left(\dfrac{h_c}{b_c}\right)(b_c - 2r_c)^2\right]}{3A_g} - \rho_g}{1 - \rho_g} \quad (9.2.4.3.1)$$

R9.2.4.2 FRP jackets are most effective at confining members with circular cross sections. The FRP system provides a circumferentially uniform confining pressure to the radial expansion of the compression member when the fibers are aligned transverse to the longitudinal axis of the member.

R9.2.4.3.1 The shape factor described by Eq. (9.2.4.3.1) depends on the aspect ratio of the cross section and the proportion of the section that is effectively confined. The accepted theoretical approach for determining A_e involves inscribing parabolic regions along each side of the section, as shown in Fig. R9.2.4.3.1. The effective confined core area of the cross section is defined by these parabolas and

CODE

9.2.5 The maximum compressive strain of FRP-confined concrete shall be determined in accordance with Eq. (9.2.5).

$$\varepsilon_{ccu} = \varepsilon_c' \left(1.5 + 12\kappa_b \frac{f_\ell}{f_c'} \left(\frac{\varepsilon_{fe}}{\varepsilon_c'}\right)^{0.45} \right) \leq 0.01 \quad (9.2.5)$$

9.2.5.1 It shall be permitted to take the value of ε_c' as 0.002.

9.2.5.2 For circular cross sections, the shape factor κ_b in Eq. (9.2.5) shall be taken as 1.0.

9.2.5.3 For rectangular cross sections having dimension $b_c \times h_c$, in which $h_c \leq b_c$, the shape factor κ_b in Eq. (9.2.5) shall be determined in accordance with Eq. (9.2.5.2).

$$\kappa_b = \frac{A_e}{A_g}\left(\frac{b_c}{h_c}\right)^{0.5} \quad (9.2.5.2)$$

9.2.5.3.1 The ratio of effective confined area of concrete to gross concrete A_e/A_g shall be determined in accordance with 9.2.4.3.1.

9.2.6 The complete stress-strain relationship for FRP-confined concrete shall be determined in accordance with Eq. (9.2.6a) through (9.2.6c).

$$f_c = \begin{cases} E_c \varepsilon_c - \dfrac{(E_c - E_2)^2}{4 f_c'} \varepsilon_c^2 & 0 \leq \varepsilon_c \leq \varepsilon_t' \\ f_c' + E_2 \varepsilon_c & \varepsilon_t' \leq \varepsilon_c \leq \varepsilon_{ccu} \end{cases} \quad (9.2.6a)$$

$$E_2 = \frac{f_{cc}' - f_c'}{\varepsilon_{ccu}} \quad (9.2.6b)$$

$$\varepsilon_t' = \frac{2 f_c'}{E_c - E_2} \quad (9.2.6c)$$

COMMENTARY

is bounded by the corner radii, r_c. ρ_g is the longitudinal steel reinforcement ratio.

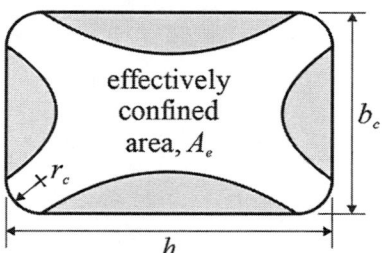

Fig. R9.2.4.3.1—Effective confined area of concrete for a rectangular section.

R9.2.5 The maximum concrete strain is limited to 0.01 to prevent excessive cracking and the resulting loss of concrete integrity. When this limit is applicable, the corresponding maximum value of f_{cc}' should be recalculated from the complete stress-strain curve given in 9.2.6.

R9.2.6 The stress-strain relationship for FRP-confined concrete is based on the model by Lam and Teng (2003a,b), as illustrated in Fig. R9.2.6. This relationship is required when establishing axial load-moment interaction relationships, for instance.

CODE

COMMENTARY

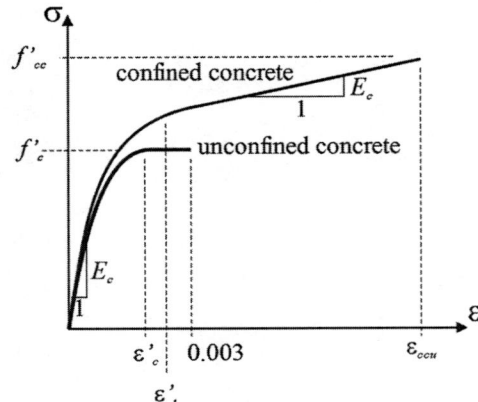

Fig. R9.2.6—Stress-strain model for FRP-confined concrete (after Lam and Teng 2003a).

9.2.7 The compressive stress in concrete under service load shall not exceed $0.60f_c'$.

R9.2.7 To ensure that radial cracking will not occur under service loads, the transverse strain in the concrete should remain below its cracking strain at service load levels. This corresponds to limiting the compressive stress in the concrete to $0.60f_c'$. By limiting the concrete stress at service load, the service load stress in the FRP jacket will be relatively low. The FRP jacket is only stressed to significant levels when the concrete is transversely strained above the cracking strain and the transverse expansion becomes large.

9.2.8 The stress in steel reinforcement under service load shall not exceed $0.60f_y$.

R9.2.8 To avoid plastic deformation under sustained or cyclic loads, the service-load-induced stress in the longitudinal reinforcing steel should remain below $0.60f_y$.

9.3—Combined axial compression and bending

9.3.1 The nominal flexural strength shall be calculated in accordance with ACI CODE-562 or Section 7.4 of this Code, as applicable.

R9.3—Combined axial compression and bending

R9.3.1 ACI CODE-562 provisions are used when no longitudinally oriented flexural strengthening is present. In structures in which longitudinally oriented FRP is present, the nominal capacity can be determined using the provisions of 7.4.

For compression-controlled sections, the confined concrete strength from Eq. (9.2.3a) or Eq. (9.2.3b), as applicable, may be used in place of the nominal compressive strength in ACI CODE-318-19 Section 22, or Section 7.4 of this Code in calculating the nominal flexural strength.

9.3.2 The nominal axial strength shall be calculated in accordance with 9.2.2.

9.3.3 The effective strain in the transversely oriented confining FRP shall not exceed 0.004 and $0.55\varepsilon_{fu}$.

R9.3.3 For predicting the effect of FRP confinement on strength enhancement, Eq. (9.2.2a) and (9.2.2b) are applicable when the load eccentricity is less than or equal to 0.1h, where h is the dimension of the member section perpendicular to the axis of bending. When eccentricity is greater than 0.1h, the methodology and equations presented in 9.2 can be used to determine the concrete material properties of the member cross section under compressive stress. Based on this, the axial load-moment (P-M) interaction diagram for

CODE

COMMENTARY

the FRP-confined member can be constructed using established procedures (Bank 2006; ACI PRC-440.2).

Strength enhancement associated with FRP confinement is more pronounced when the applied ultimate axial force P_u and bending moment M_u fall above the line connecting the origin and the balance point of the P-M diagram (Fig. R9.3.3). P-M diagrams may be developed by satisfying strain compatibility and force equilibrium using the model for the stress-strain behavior for FRP-confined concrete presented in 9.2.5. For simplicity, the portion of the unconfined and confined P-M diagrams corresponding to compression-controlled failure can typically be reduced to two bilinear curves passing through three points, as presented in Fig. R9.3.3:

Point A (pure compression) is determined from 9.2.

Point B corresponds to a sectional strain distribution corresponding to zero strain at the layer of longitudinal steel reinforcement nearest to the tensile face, and a compressive strain ε_{ccu} at the compression face of the member.

Point C corresponds to a sectional strain distribution corresponding balanced failure with a maximum compressive strain ε_{ccu} and a yielding tensile strain ε_{sy} at the layer of longitudinal steel reinforcement nearest to the tensile face.

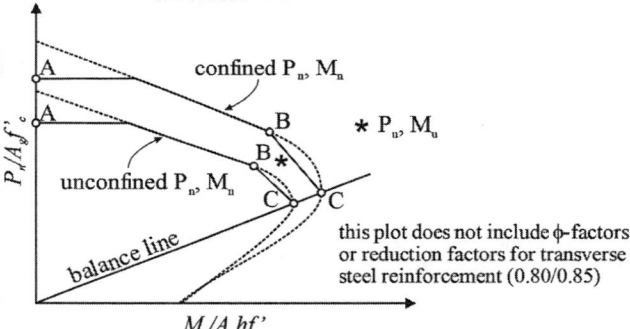

Fig. R9.3.3—Representative P-M interaction diagram.

CODE

CHAPTER 10—FIRE RESISTANCE

10.1—General

10.1.1 This chapter provides requirements for determining the fire resistance of concrete members strengthened with externally bonded FRP systems.

10.1.2 FRP-strengthened concrete elements shall be able to resist the effects of gravity loads during a fire event in accordance with 6.3.3.

10.2—Fire resistance of FRP-strengthened members

10.2.1 Fire resistance of FRP-strengthened concrete structures shall be determined following the requirements of the design basis code in accordance with ACI CODE-562.

10.2.2 The contribution of the FRP strengthening system shall be neglected for purposes of calculating strength during a fire event, regardless of the use of fire protection.

10.2.3 Fire resistance shall be determined by analysis in accordance with 10.2.3.1 or by qualification testing in accordance with 10.2.3.2.

COMMENTARY

CHAPTER R10—FIRE RESISTANCE

R10.1—General

R10.1.1 The requirements in this chapter are limited to determining structural fire resistance during a fire event. Following exposure to a fire, the condition of the FRP, concrete, and internal steel reinforcement should be evaluated to determine the residual strength of affected members. ACI CODE-562 provides guidance on the assessment of concrete structures following a fire event. This chapter does not cover all requirements relating to fire that may be mandated by the design basis code, such as flame spread and smoke generation characteristics as evaluated by ASTM E84.

R10.1.2 Code requirements for determining fire resistance of strengthened concrete members are given in ACI CODE-562. Guidance specific to FRP-strengthened members is available in ACI PRC-440.10. A fire event is considered an extreme event and should not be combined with other extreme events such as an earthquake or a hurricane.

The degree of strengthening that can be achieved using externally bonded FRP can be indirectly limited by the code-required fire-resistance rating of a structure. The physical and mechanical properties of the resin components of FRP systems are influenced by temperature and degrade at temperatures close to or above their glass transition temperature T_g. Additional discussion is found in ACI PRC-440.2. Although the FRP system itself is affected by exposure to elevated temperature, an existing concrete structure can still have an adequate fire resistance.

R10.2—Fire resistance of FRP-strengthened members

R10.2.2 The strength contribution of adhesively bonded external FRP reinforcing systems, including near-surface-mounted (NSM) FRP systems, during a fire event is not well established. To address this, the strength contribution of FRP reinforcement is neglected for strength calculations during a fire event. Nevertheless, fire protection systems can be used with FRP-strengthened members to improve the fire resistance of the existing concrete member.

R10.2.3 Determining the fire resistance of FRP-strengthened concrete members follows the same approach used to for new members; however, the strength contribution of the FRP system is neglected. The strength of a reinforced concrete element is reduced during fire exposure due to increase in temperature of both the reinforcing steel and the concrete. ACI CODE-216.1 provides methods for determining the fire resistance of concrete structural elements based on member geometry, materials, and reinforcement details; the effect

CODE	COMMENTARY
	of the presence of FRP on temperature profiles within the member is not likely to be significant. The performance of concrete elements due to fire exposure depends on the temperature distribution within the elements, the maximum temperature reached during fire, and the resulting reduction of mechanical properties of steel and concrete due to fire exposure. If fire protection systems (that is, thermal insulation) are present, their effects on material strengths during fire exposure should be considered. The strength contribution of the FRP system during a fire event should be neglected, regardless of the use of fire protection.
10.2.3.1 Analytical methods to determine the fire resistance of FRP-strengthened concrete members based on temperature profile within the concrete section shall be in accordance with ACI CODE-216.1. Alternative methods of determining temperature profiles shall be permitted if based on principles of heat transfer.	**R10.2.3.1** An alternative method of determining temperature profiles is finite-element-based thermal modeling of the concrete element for a specific time-temperature curve. The strength of members at any time during a fire event is based on reduced material properties at the elevated temperatures.
10.2.3.2 Qualification of the fire resistance of FRP-strengthened assemblies using results of furnace tests shall be in accordance with the procedures of ASTM E119 or ANSI/UL 263.	**R10.2.3.2** Qualification testing provides prescriptive details of the assembly and restraint conditions tested. Results of ASTM E119 furnace tests to demonstrate adequate fire resistance are most applicable to FRP applications that have identical characteristics to the tested assembly. Deviations from the tested assembly that affect thermal performance, such as member geometry, concrete cover, and strengthening ratio, should be evaluated by an experienced design professional to evaluate the applicability of a qualified design to project-specific applications. The fire rating achieved by furnace tests applies only to the entire test assembly. Fire ratings are not assigned to fire protection systems (that is, thermal insulation), if present, or other individual components of the test assembly.
10.2.4 Improving the fire resistance of FRP-strengthened concrete members using externally applied fire protection shall be permitted.	**R10.2.4** Externally applied fire protection systems include spray applied materials, intumescent coatings, an additional layer of concrete or cementitious materials, gypsum boards, and other materials providing thermal insulation. Externally applied fire protection can improve the fire resistance by reducing the temperature of the member during fire and, consequently, the adverse effects on existing concrete and internal steel reinforcement. Few fire protection systems are able to maintain the FRP temperature below its glass transition temperature (Bisby et al 2005; Palmieri et al. 2011; ACI PRC-440.2). Fire protection systems are typically tested as part of a test assembly based on ASTM E119 or ANSI/UL 263, or as required by the authority having jurisdiction. Analytical evaluation of FRP-strengthened concrete members with fire protection requires knowledge of the thermal properties of the insulation material, including thermal conductivity, specific heat, and density at elevated temperatures, which can be determined in accordance with ASTM C1113/C1113M or ISO 22007-2. In addition, the ability of the insulation material to remain thermally effective and adequately bonded to the substrate during the fire event should be understood and substantiated by tests.

CODE

CHAPTER 11—FIELD INSPECTION, TESTING, AND EVALUATION

11.1—General

11.1.1 Field inspection and evaluation of installed FRP systems shall be performed as required by the governing building code and as specified in the construction documents.

11.2—Field inspection

11.2.1 The construction documents shall contain job-specific and building-code-compliant descriptions of required inspections, tests, evaluations, and acceptance criteria.

COMMENTARY

CHAPTER R11—FIELD INSPECTION, TESTING, AND EVALUATION

R11.1—General

R11.1.1 This chapter covers only quality-assurance activities conducted during installation. Quality-control activities performed prior to installation generally consist of requiring the manufacturer and contractor to submit product information and evidence of qualifications to the licensed design professional responsible for the construction documents for review.

In jurisdictions where the International Building Code (IBC-2021) is the governing code, Section 1705.1 should apply for FRP systems because such systems are considered as an alternate to IBC-prescribed materials and systems. In addition, IBC-2021 Section 1704.3 requires a statement of special inspections prepared by the design professional in responsible charge. ACI CODE-562 also contains requirements for inspection and related quality assurance measures that may be applicable.

R11.2—Field inspection

R11.2.1 Procedures for installing FRP systems have been developed by the system manufacturers and often differ between systems. In addition, surface preparation and installation procedures can vary within a system, depending on the type and condition of the structure. Deviations from the procedures developed by the FRP system manufacturer should not be permitted without consulting with the licensed design professional and the FRP system manufacturer. Because FRP installations vary significantly with respect to scope, type, structural demands, application criteria, regional practice, and other factors, this Code can only provide general guidance regarding common inspection requirements. The licensed design professional should review IBC-2021, ACI PRC-440.2, ACI CODE-562, ICRI 330.2, and ICC-ES AC178 for additional guidance.

The following is a nonexhaustive list of items commonly included in a field inspection program:

(a) Materials: Verification that the specified FRP system is being installed

(b) Concrete substrate: Inspection of general condition, including moisture condition, treatment of corners, protrusions, cracks, deteriorated concrete, corrosion of internal steel reinforcement, interfering embedments, and surface contaminants

(c) Surface preparation: Inspection of preparation method and resulting profile; or NSM systems, inspection of preparation, dimensions, and cleaning of groove

(d) Recording of environment at the time of installation of the FRP system including air and surface temperature, humidity, concrete moisture, and dew point

(e) Inspection of resin: For wet lay-up systems, resin mixing and application to fiber; for precured laminates, adhesive mixing and application to substrate or laminate.

(f) Verification of relative resin cure

CODE

COMMENTARY

(g) Inspection of fiber placement: Fiber layout, dimensions, spacing, number of layers, splices, and fiber direction and alignment

(h) For NSM systems, sample cores may be extracted to visually assess the degree of consolidation of the resin adhesive around the FRP bar

(i) Verification of fiber anchors used to anchor U-wrap shear reinforcement, including anchor weight per unit length, fiber and epoxy resin properties, number and spacing of FRP anchors, fiber splay length and width, embedment depth, drilled hole diameter and hole drilling, and cleaning procedures, as applicable; pull-out testing may be used to verify as-installed anchor capacity

(j) Verification of pull-off strength in accordance with ASTM D7522/D7522M for externally bonded systems and in accordance with ACI SPEC-440.12 for wet-layup systems

(k) Inspection of witness panels for wet-layup systems

(l) Inspection of delaminations: Acoustic sounding, infrared thermography, or other diagnostic measures capable of detecting delaminations 2 in.^{2} or smaller in cured FRP system

(m) Inspection of remedial measures: Inspection of epoxy injection of delaminations, reinstallation of unacceptably installed fibers, or other remedial measures deemed necessary by the licensed design professional

11.2.2 The extent and frequency of inspections shall be determined by the licensed design professional based on job-specific parameters.

11.2.3 Written reports documenting field observations and pull-off strength test results shall be prepared by the inspector and submitted to appropriate parties. The reports shall document the date and time of the inspection, installation location on structure, batch numbers, resin/adhesive mixture ratios and mixing times, and other general information.

R11.2.3 Inclusion of annotated plans showing the location of inspected work and photographs to illustrate special conditions will clarify the report narrative. All deviations from the drawings, specifications, or manufacturer's instruction should be clearly identified and promptly communicated to appropriate parties.

11.3—Material testing

11.3.1 FRP materials shall be evaluated for compliance with properties reported by the manufacturer in accordance with 11.3.1.1, 11.3.1.2, 11.3.1.3, or 11.3.1.4.

11.3.1.1 Evaluation of a wet layup system shall be performed by fabricating and testing witness panels in accordance with ACI SPEC-440.12. Values shall be reported as required by 4.2.

11.3.1.2 For precured carbon FRP plate systems, the licensed design professional shall require the manufacturer to submit test results obtained in accordance with ASTM D7565/D7565M or ASTM D3039/D3039M. Values shall be reported as required by 4.3.

11.3.1.3 For glass NSM FRP, the licensed design professional shall require the manufacturer to submit test results

CODE

obtained in accordance with ASTM D7957/D7957M. Values shall be reported as required by 4.4.1.

11.3.1.4 For carbon NSM FRP, the licensed design professional shall require the manufacturer to submit test results obtained in accordance with ACI SPEC-440.6. Values shall be reported as required by 4.4.2.

11.3.2 Evaluation of adhesives used for precured externally applied FRP systems and NSM systems shall be performed and reported in accordance with 4.3.5.

11.4—Evaluation and acceptance criteria

11.4.1 The construction documents shall set forth evaluation and acceptance criteria so that material properties, quality, and workmanship expectations are clearly communicated to the contractor.

11.4.1.1 For wet layup FRP systems, evaluation, installation tolerance and acceptance criteria shall be in accordance with ACI SPEC-440.12.

11.4.2 The effect of deviations from installation tolerances on the structural performance of the FRP system shall be reviewed by the licensed design professional.

11.4.3 The effect of delaminations on the structural performance and durability of the FRP system shall be reviewed by the licensed design professional based on the size, location, quantity, and other factors.

11.5—Inspection of coatings

11.5.1 Protective coatings in accordance with 6.7.3 or 10.2.4 shall be inspected as required by the construction documents, governing building code requirements for special inspection, and product specific evaluation service report.

COMMENTARY

STRENGTHENING STRUCTURAL CONCRETE WITH FRP SYSTEMS (ACI CODE-440.13-24)

CODE	COMMENTARY
	COMMENTARY REFERENCES *American Concrete Institute (ACI)* ACI CODE-369.1-22—Seismic Evaluation and Retrofit of Existing Concrete Buildings—Code and Commentary ACI CODE-216.1-14—Code Requirements for Determining Fire Resistance of Concrete and Masonry Construction Assemblies ACI CODE-318-19(22)—Building Code Requirements for Structural Concrete and Commentary ACI PRC-440.2-23—Guide for the Design and Construction of Externally Bonded FRP Systems for Strengthening Concrete Structures ACI SPEC-440.6-08(22)—Specification for Carbon Fiber-Reinforced Polymer Bar Material for Concrete Reinforcement ACI PRC-440.7-22—Externally Bonded Fiber-Reinforced Polymer Systems Design and Construction for Strengthening Masonry Structures—Guide ACI SPEC-440.8-13(23)—Specification for Carbon and Glass Fiber-Reinforced Polymer Materials Made by Wet Layup for External Strengthening ACI PRC-440.10-21—Fire Resistance of FRP-Strengthened Concrete Members—TechNote ACI SPEC-440.12-22—Specification for Strengthening of Concrete Structures with Externally Bonded Fiber-Reinforced Polymer (FRP) Materials using the Wet Layup Method ACI PRC-546-14—Guide to Concrete Repair ACI CODE-562-21—Assessment, Repair, and Rehabilitation of Existing Concrete Structures—Code and Commentary *American Society of Civil Engineers (ASCE)* ASCE/SEI 7-22—Minimum Design Loads and Associated Criteria for Buildings and Other Structures ASCE/SEI 41-17—Seismic Evaluation and Retrofit of Existing Buildings *ASTM International* ASTM C1113/C1113M-09(2019)—Standard Test Method for Thermal Conductivity of Refractories by Hot Wire (Platinum Resistance Thermometer Technique). ASTM D3039/3039M-17—Standard Test Method for Tensile Properties of Polymer Matrix Composite Materials ASTM D4258-05(2017)—Standard Practice for Surface Cleaning Concrete for Coating ASTM D7522/D7522M-15—Standard Test Method for Pull-Off Strength for FRP Bonded to Concrete Substrate ASTM D7565/D7565M-10(2017)—Standard Test Method for Determining Tensile Properties of Fiber Reinforced Polymer Matrix Composites Used for Strengthening of Civil Structures ASTM D7682-17—Standard Test Method for Replication and Measurement of Concrete Surface Profiles Using Replica Putty ASTM D7957/D7957M-22—Standard Specification for Solid Round Glass Fiber Reinforced Polymer Bars for Concrete Reinforcement

CODE

COMMENTARY

ASTM E84-23—Standard Test Method for Surface Burning Characteristics of Building Materials

ASTM E119-22—Standard Test Methods for Fire Tests of Building Construction and Materials

ICC Evaluation Services

ICC-ES AC178:2020—Inspection and Verification of Concrete and Reinforced and Unreinforced Masonry Strengthening Using Fiber-Reinforced Polymer (FRP) and Steel-Reinforced Polymer (SRP) Composite Systems

International Code Council

IBC-2021—International Building Code

International Concrete Repair Institute (ICRI)

ICRI 310.2R-2013—Selecting and Specifying Concrete Surface Preparation for Sealers, Coatings, Polymer Overlays, and Concrete Repair

ICRI 330.2-2016—Guide Specifications for Externally Bonded FRP Fabric Systems for Strengthening Concrete Structures

International Organization for Standardization (ISO)

ISO 22007-2:2017—Plastics – Determination of Thermal Conductivity and Thermal Diffusivity - Transient Plane Heat Source (Hot Disc) Method

Precast/Prestressed Concrete Institute (PCI)

PCI MNL-120-17—PCI Design Handbook

Underwriters Laboratories

ANSI/UL 263-11—Fire Tests of Building Construction and Materials

Authored documents

Bank, L. C., 2006, *Composites for Construction: Structural Design with FRP Materials*, John Wiley & Sons, Hoboken, NJ, 560 pp.

Bisby, L. A.; Green, M. F.; and Kodur, V. K. R., 2005, "Fire Endurance of Fiber-Reinforced Polymer-Confined Concrete Columns," *ACI Structural Journal*, V. 102, No. 6, Nov.-Dec., pp. 883-891.

Cromwell, J. R.; Harries, K. A.; and Shahrooz, B. M., 2011, "Environmental Durability of Externally Bonded FRP Materials Intended for Repair of Concrete Structures," *Construction and Building Materials*, V. 25, No. 5, pp. 2528-2539. doi: 10.1016/j.conbuildmat.2010.11.096

del Rey Castillo, E.; Ingham, J. M.; Smith, S. T.; Kanitkar, R. V.; and Griffith, M. C., 2019, "Design Approach for FRP Spike Anchors in FRP-strengthened RC Structures," *Composite Structures*, V. 214, pp. 23-33. doi: 10.1016/j.compstruct.2019.01.100

El Meski, F., and Harajli, M., 2014, "Flexural Capacity of Fiber-Reinforced Polymer Strengthened Unbonded Post-Tensioned Members," *ACI Structural Journal*, V. 111, No. 2, Mar.-Apr., pp. 407-418. doi: 10.14359/51686565

CODE

COMMENTARY

El-Refaie, S. A.; Ashour, A. F.; and Garrity, S. W., 2003, "Sagging and Hogging Strengthening of Continuous Reinforced Concrete Beams using Carbon Fiber-Reinforced Polymer Sheets," *ACI Structural Journal*, V. 100, No. 4, July-Aug., pp. 446-453. doi: 10.14359/12653

Ellingwood, B., 2005, "Load Combination Requirements for Fire-Resistant Structural Design," *Journal of Fire Protection Engineering*, V. 15, No. 1, pp. 43-61. doi: 10.1177/1042391505045

Eshwar, N.; Nanni, A.; and Ibell, T. J., 2005, "Effectiveness of CFRP Strengthening on Curved Soffit RC Beams," *Advances in Structural Engineering*, V. 8, No. 1, pp. 55-68. doi: 10.1260/1369433053749607

Grace, N. F., and Singh, S. B., 2005, "Durability Evaluation of Carbon Fiber-Reinforced Polymer Strengthened Concrete Beams: Experimental Study and Design," *ACI Structural Journal*, V. 102, No. 1, Jan.-Feb., pp. 40-53. doi: 10.14359/13529

Harajli, M., 2012, "Tendon Stress at Ultimate in Continuous Unbonded Post-Tensioned Members: Proposed Modification to ACI Eq. (18-4) and (18-5)," *ACI Structural Journal*, V. 109, No. 2, Mar.-Apr., pp. 183-192.

Jirsa, J. O.; Ghannoum, W. M.; Kim, C.; Sun, W.; Shekarchi, W.; Alotaibi, N.; Pudleiner, D. K.; Zhu, J.; Liu, S.; and Wang, H., 2017, "Use of Carbon Fiber Reinforced Polymer (CFRP) with CFRP Anchors for Shear-Strengthening and Design Recommendations/Quality Control Procedures for CFRP Anchors," *FHWA/TX-16/0-6783-1*, Center for Transportation Research (CTR), Mar. 278 pp.

Lam, L., and Teng, J., 2003a, "Design-Oriented Stress-Strain Model for FRP-Confined Concrete," *Construction and Building Materials*, V. 17, No. 6-7, pp. 471-489. doi: 10.1016/S0950-0618(03)00045-X

Lam, L., and Teng, J., 2003b, "Design-Oriented Stress-Strain Model for FRP-Confined Concrete in Rectangular Columns," *Journal of Reinforced Plastics and Composites*, V. 22, No. 13, pp. 1149-1186. doi: 10.1177/0731684403035429

Lopez, M. M.; Naaman, A. E.; Pinkerton, L.; and Till, R. D., 2003, "Behavior of RC Beams Strengthened with FRP Laminates and Tested under Cyclic Loading at Low Temperatures," *International Journal of Materials & Product Technology*, V. 19, No. 1, pp. 108-117. doi: 10.1504/IJMPT.2003.003549

Palmieri, A.; Matthys, S.; and Taerwe, L., 2011, "Fire Testing of RC Beams Strengthened with NSM Reinforcement," 10th International Symposium on Fiber-Reinforced Polymer Reinforcement for Concrete Structures (FRPRCS-10), SP-275, American Concrete Institute, Farmington Hills, MI. (CD-ROM).

Pudleiner, D. K.; Jirsa, J. O.; and Ghannoum, W. M., 2019, "Influence of Anchor Size on Anchored CFRP Systems," *Journal of Composites for Construction*, ASCE, V. 23, No. 5, p. 04019033. doi: 10.1061/(ASCE)CC.1943-5614.0000954

Rocca, S.; Galati, N.; and Nanni, A., 2006, "Experimental Evaluation of FRP Strengthening of Large-Size Reinforced Concrete Columns," *Report* No. UTC-142, University of Missouri-Rolla, MO.

CODE

APPENDIX A—ADDITIONAL LOAD COMBINATIONS FOR FRP STRENGTHENING

A.1—Appendix notation
D = dead load effects
L = live load effects
R_{ex} = nominal strength of a member subject to fire or elevated temperature calculated using reduced material properties
R_n = nominal strength of a member without the contribution of the FRP strengthening system
S = snow load effects
ϕ = strength reduction factor
ϕ_{ex} = strength reduction factor used to check strength of member subject to fire or elevated temperature calculated using reduced material properties

A.2—Scope
A.2.1 This Appendix supplements the requirements of Section 6.3.

A.3—Additional load combinations for FRP strengthening
A.3.1 The strength of the concrete element without FRP strengthening should satisfy Eq. (A.3.1a) and (A.3.1b).

$$\phi R_n \geq 1.1D + 0.5L + 0.2S \quad (A.3.1a)$$

$$\phi R_n \geq 1.1D + 0.75L \quad (A.3.1b)$$

where D, L, and S are the effects due to the dead, live, and roof snow loads, respectively, calculated for the strengthened structure; ϕ is the strength reduction factor used for design; and R_n is the nominal strength of the structural member without the contribution of the FRP strengthening system.

A.3.2 If a fire-resistance rating is required by the design basis code, FRP-strengthened concrete members should satisfy Eq. (A.3.2).

$$\phi_{ex} R_{ex} \geq (0.9 \text{ or } 1.2)D + 0.5L + 0.2S \quad (A.3.2)$$

where $\phi_{ex} = 1.0$; R_{ex} is the nominal strength of the structural member at elevated temperature without the contribution of the FRP strengthening system and considering reduced concrete and steel material strengths due to fire exposure.

COMMENTARY

APPENDIX RA—ADDITIONAL LOAD COMBINATIONS FOR FRP STRENGTHENING

A.2—Scope
RA.2.1 The information provided in this appendix is consistent with the load combinations given in ACI CODE-562-21 Sections 5.5.2 and 5.5.3, pertaining to members strengthened with FRP, and is provided as a convenience to the user.

RA.3—Additional load combinations for FRP strengthening
RA.3.1 This load combination is intended to minimize the risk of failure of the strengthened structural member in the case where, during normal operating conditions, the external reinforcement is damaged. If such damage is not detected immediately, the ability of the structure (or component) to resist full design loads may be compromised until the damage is identified and addressed. The live load factor of 0.75 is used in Eq. (A.3.1b) to exceed the statistical mean of the yearly maximum live load factor of 0.5, given in ASCE/SEI 7. Effects of: a) internal load effects due to reactions induced by prestressing; b) fluid loads; and c) lateral earth pressure should be included in Eq. (A.3.1a) and (A.3.1b) with a factor of 1.0 when the effect of these loads acts alone or adds to the primary load effect.

If the live load has a high likelihood of being a sustained load (for example, library stack areas, heavy storage areas, warehouses, and other storage occupancies with a live load exceeding 100 lb/ft^2), the live load factor in Eq. (A.3.1a) and (A.3.1b) should be increased to 1.0.

RA.3.2 Equation (A.3.2) is intended to ensure that the repaired element will maintain sufficient strength, accounting for its probable reduced material properties due to elevated temperatures, during a fire event. This load combination is representative of a typical loading condition during a fire event (Ellingwood 2005). The strength of the member, R_{ex}, should be based on reduced material properties due to exposure to elevated temperatures during a fire event as described in Chapter 10 of this Code. If fire protection is applied to the existing or strengthened member, its effect on

CODE

The dead load factor of 0.9 is applied only when the dead load effect counteracts the total load effect.

COMMENTARY

the existing concrete elements and existing reinforcement should be considered in determining the nominal resistance of the member. The strength contribution of the FRP system during a fire event should be neglected, regardless of the use of fire protection.

If the live load has a high likelihood of being a sustained load (for example, library stack areas, heavy storage areas, warehouses, and other storage occupancies with a live load exceeding 100 lb/ft^2), the live load factor in Eq. (A.3.2) should be increased to 1.0.